# THE LIFE OF THE CITY

# Studies in Historical Geography

*Series Editor*: Professor Robert Mayhew, University of Bristol, UK

Historical geography has consistently been at the cutting edge of scholarship and research in human geography for the last fifty years. The first generation of its practitioners, led by Clifford Darby, Carl Sauer and Vidal de la Blache presented diligent archival studies of patterns of agriculture, industry and the region through time and space.

Drawing on this work, but transcending it in terms of theoretical scope and substantive concerns, historical geography has long since developed into a highly interdisciplinary field seeking to fuse the study of space and time. In doing so, it provides new perspectives and insights into fundamental issues across both the humanities and social sciences.

Having radically altered and expanded its conception of the theoretical underpinnings, data sources and styles of writing through which it can practice its craft over the past twenty years, historical geography is now a pluralistic, vibrant and interdisciplinary field of scholarship. In particular, two important trends can be discerned. Firstly, there has been a major 'cultural turn' in historical geography which has led to a concern with representation as driving historical-geographical consciousness, leading scholars to a concern with text, interpretation and discourse rather than the more materialist concerns of their predecessors. Secondly, there has been a development of interdisciplinary scholarship, leading to fruitful dialogues with historians of science, art historians and literary scholars in particular which has revitalised the history of geographical thought as a realm of inquiry in historical geography.

Studies in Historical Geography aims to provide a forum for the publication of scholarly work which encapsulates and furthers these developments. Aiming to attract an interdisciplinary and international authorship and audience, Studies in Historical Geography will publish theoretical, historiographical and substantive contributions meshing time, space and society.

# The Life of the City
## Space, Humour, and the Experience of Truth in Fin-de-siècle Montmartre

JULIAN BRIGSTOCKE
*Plymouth University, UK*

Routledge
Taylor & Francis Group

LONDON AND NEW YORK

First published 2014 by Ashgate Publishing

Published 2016 by Routledge
2 Park Square, Milton Park, Abingdon, Oxfordshire OX14 4RN
711 Third Avenue, New York, NY 10017, USA

First issued in paperback 2016

*Routledge is an imprint of the Taylor & Francis Group, an informa business*

**British Library Cataloguing in Publication Data**
A catalogue record for this book is available from the British Library.

**The Library of Congress has cataloged the printed edition as follows:**
Brigstocke, Julian.
    The life of the city : space, humour and the experience of truth in fin de siecle Montmartre / by Julian Brigstocke.
        pages cm. – (Studies in historical geography)
    Includes bibliographical references and index.
    ISBN 978-1-4094-4896-9 (hardback)
  1. City planning – France – Paris – History. 2. Urban ecology (Sociology) – France – Paris – History. 3. Montmartre (Paris, France) – Social conditions. I. Title.

    HT169.F72M656 2014
    307.1'2160944361–dc23

                                                                                2013048256

ISBN 13: 978-1-138-25077-2 (pbk)
ISBN 13: 978-1-4094-4896-9 (hbk)

# Contents

# Contents

# List of Figures

# Preface

A young woman stands over the body of a slaughtered cow.[1] Young, pretty and innocent looking, she calmly wipes a streak of crimson blood off her sword onto her ghoulishly black dress. The ribbons of the Tricolore flutter from the lance in the cow's side. With a petite foot extending over the dead beast's enormous, distended eye, the woman offers the viewer a flirtatious glimpse of ankle and petticoat. Behind her lies the entrance to Montmartre, the fin-de-siècle centre of French revolutionary counter-culture; at the top of the hill we see the windmills for which it is famous, as well as the construction site of the Sacré Coeur Basilica, that symbol of the worst degradation of the modern metropolis, the violent suppression of the Paris Commune and the egalitarian rights to the city that it proclaimed. The image is full of obvious contrasts: black and red; innocence and death; rebellion and reaction; humour and violence. The cow she has killed is the 'vache enragée', a metaphor for hunger, poverty, and abjection.[2] It is from Montmartre, the image proclaims, that a true Republic will rise up, one which does not exist to protect the cynical modernity of bourgeois wealth-extraction, but to defend a new set of democratic values. Equality. Imagination. Pleasure.

Today, the winding streets of Montmartre are filled by an endless flow of tourists looking to enjoy the views over Paris and to take in the area's atmosphere of quaint creativity and bohemian lifestyle. A potent 'realm of memory', Montmartre has come to embody a nostalgia for a romanticized image of liberal freedom – a cult of creativity, aestheticized individualism, and bohemian chic.[3] The quirkiness of *Amélie*; the decadence of *Moulin Rouge*.[4] It has become a kind of symbol of urban vitality and dynamism – albeit one that is divested of any sense of political engagement or radical critique. The Paris Commune, the anarchist plots, the feverish rumours of piles of dynamite lurking within the foundations of the Sacré Coeur – these have all but disappeared from the sanitized memory

---

1 See the front cover of this book.

2 On the *vache enragée*, see Chapter 5.

3 Nicholas Hewitt, 'From 'Lieu De Plaisir' to 'Lieu De Mémoire': Montmartre and Parisian Cultural Topography', *French Studies* LIV(4) (2000). On realms of memory, see Pierre Nora (ed.), *Realms of Memory: The Construction of the French Past*, 3 vols (New York, 1998). On the role of 'creativity' in supporting structures of contemporary liberal government, see Thomas Osborne, 'Against 'Creativity': A Philistine Rant', *Economy and Society* 32(4) (2003).

4 Jean-Pierre Jeunet (dir.), *Le Fabuleux Destin D'Amélie Poulain*, produced by Jean-Marc Deschamps and Claudie Ossard (2001), Baz Luhrmann (dir.), *Moulin Rouge!*, produced by Fred Baron, Martin Brown, and Baz Luhrmann (2001).

of what Montmartre represented during its bohemian heyday. Forgotten too are Montmartre landscapes that were fought for but never materialized: a vast Statue of Liberty (just like the one then being built in Paris for export to New York), for example, that was to be placed directly in front of the Sacré Coeur to disrupt the Basilica's command over the Parisian skyline.[5]

The origins of Montmartre's transformation from a neglected, poverty-stricken suburb of Paris into a centre of French counter-culture, however, can be found in the experiments of an urban avant-garde that was closely allied to anarchist ideals of local autonomy, self-determination, and equality. Poets, musicians and artists associated with small 'cabarets artistiques' in Montmartre aimed to invent an urban ethos that would be radically opposed to the established values of fin-de-siècle liberalism. Anticipating later urban avant-gardes such as Surrealism and Situationism, the literary and artistic experiments of groups associated with cabarets such as the Chat Noir, infused with an anarchic vitality and sense of humour, attempted to actualize a new urban geographical imagination and a transformed art of urban living.

This book is a contribution to a spatial history of this art of urban life.[6] It advances scholarly debates concerning: the spatial politics of affect and, more broadly, the politics of aesthetics; culture and biopolitical experience; spatialities of the urban avant-garde; the genealogy of non-representational spatial practices and politics; the role of humour in the production and contestation of space and place; and the intersections of humour and violence in anarchist urban politics. Its principal theoretical innovation to these fields is to frame the cultural politics of the late nineteenth century in terms of a struggle for *authority*. Rather than accepting the widespread narrative that frames avant-garde urban culture and anarchist cultural politics in terms of a refusal of authority, I will argue that these groups attempted to deploy new forms of 'immanent' authority that were based on the vital energies of embodied experience.

*The Life of the City* is a study of those artistic and political activities in Montmartre that were devoted to creatively altering the limits of and relations between life, authority, and urban experience. Combining archival research, textual analysis and original theoretical work, the book turns to a period in which the life of the city was starting to be problematized in a way that was distinctively modern. It was during the last decades of the nineteenth century that the roots of modern urbanism, involving an ideal of the planned city as a regulator of modern society, and an aim to transform the socio-natural milieu into a healthy and stable environment, were planted.[7] It was also during this period that the notion of a cultural and political urban avant-garde started to emerge. The book charts

---

5    David Harvey, 'Monument and Myth', *Annals of the Association of American Geographers* 69(3) (1979), p. 379.

6    On spatial history, see Stuart Elden, *Mapping the Present: Heidegger, Foucault and the Project of a Spatial History* (London, 2001).

7    Paul Rabinow, *French Modern: Norms and Forms of the Social Environment* (Cambridge, MA, 1989).

aspects of the urban culture of 1880s Montmartre, and in doing so, challenges an influential social-scientific narrative in which the rich history of nineteenth-century urban culture is reduced to a story of ongoing 'spectacularization', abstraction, and aestheticization of everyday life. In the narrative which the book contests, embodied experience becomes reduced to abstract discourse, active citizens are reduced to passive spectators, and life is reduced to representation.

The biopolitical urban culture of 1880s Montmartre cannot usefully be understood in such terms. By viewing late nineteenth-century French culture, not in terms of an economy of representation, but in terms of an economy of life and lived experience, tired analytical contrasts between popular and avant-garde urban culture can be replaced by a more productive set of distinctions. Chief amongst these is a contrast between practices that aim to proliferate life and lived experience by framing them within existing hegemonic knowledges about true healthy life, and practices that seek to find a way to life's *outside*, the leakages beyond those discourses and assemblages of power through which life is controlled and delimited. This distinction opens up a set of questions which bring a novel perspective to the heavily researched field of fin-de-siècle Parisian urban culture.

First, the book offers a novel analysis of the 'biopolitical' roots of the cultural and political avant-gardes of the late nineteenth-century European city. The avant-garde is often portrayed as an attempt to collapse the distinction between art and life by taking creativity to the streets and styling everyday life as a work of art. Yet acknowledging the role of biopolitical discourses in the idea of the avant-garde requires us to take seriously the ways in which the converse was also true; in other words, art was also to be styled as a form of life. Advancing important debates concerning the ways in which biological life and affective, nonrepresentational modes of experience can be a source of resistance to or evasion of power, as well as being the object of government, the book unravels some of the political ambiguities and contradictions of cultural biopolitics.[8] In contrast to the usual focus on the role of technical expertise in governing the life of the city, the book demonstrates that the arts were also an important vehicle for embodied experiments with the nature of urban vitality in the modern city. In doing so, the book contributes to the historical geography of 'non-representational' spatial practices.

Second, the book deploys a new analytical framework concerning the relationship between culture, life and authority. In contrast to accounts of avant-garde urban culture that portray it as having been inherently anti-authoritarian, I argue instead that the urban culture of fin-de-siècle Montmatre attempted to invent a new form of authority, one based on the authority of affective experience rather than the authority of tradition or scientific expertise. Thus I open up an analytical approach towards reconceptualizing urban cultural politics in terms of the invention of new

---

8    On affect and biopolitics see, for example, Ben Anderson, 'Affect and Biopower: Towards a Politics of Life', *Transactions of the Institute of British Geographers* 37(1) (2012), Brian Massumi, *Parables for the Virtual: Movement, Affect, Sensation* (Durham, NC, 2002), Nigel Thrift, *Non-Representational Theory: Space, Politics, Affect* (London, 2007).

structures of authority following the 'crisis of authority' associated with the late nineteenth and early twentieth centuries.

Finally, this is the first book-length study to tackle the role of humour in the production and contestation of urban space. Humour, I argue, was a powerful technique for creating a sense of place that was capable of encompassing the contradictions associated with modern living. Moreover, as a particularly powerful form of affective and biological life, humour emerged as an important device for contesting dominant biopolitical knowledges concerning the nature of true urban health and vitality. Developing this line of analysis, the book also draws out the surprising inter-relations between humour and violence in the modern city and, in particular, in late nineteenth-century anarchist urban politics.

## Book Outline

The book is divided into three parts. Part I, 'The Life of the City', outlines the main themes of the book. Chapter 1 introduces the urban counter-cultures associated with fin-de-siècle Montmartre and their attempts to invent new relationships between art, life and experience. It goes on to discuss the role of life and lived experience in the art of life in biopolitical modernity, and elaborates a critique of the narrative of the 'spectacular' city that dominates much writing on the nineteenth-century urban experience and the historical avant-garde. It argues that if we are to fully understand the role of biopolitical discourses in the constitution of modern urban experience, we must conduct an investigation of those forms of experience that tested and contested the limits of the experiential life of the city. Scholarly accounts of the biopolitical city, preoccupied with the role of technical expertise in the government of the city, have not done enough to analyze the important role of the arts in experimenting with new knowledges and experiences of urban vitality and liveliness.

Perhaps the most important symbolic figure of fin-de-siècle Montmartre culture was the black cat, after which Montmartre's most famous cabaret, the Chat Noir, was named. Chapter 2, through an analysis of works by Poe, Manet and Baudelaire, explores the ways in which the black cat emerged as an important figure of the experience of truth in modern French culture. In these works, truth was portrayed as something that cannot be represented but only encountered through absence. Truth was not to be found in the interior of life and lived experience, but in what lay beyond life. Via a discussion of Michel Foucault's remarks on the ancient Cynics, the chapter argues that the black cat presents a particular figure of truth-telling: one where accessing the truth requires a creative, artistic transformation of life and the limits of lived experience. In the modern city, I suggest, accessing truth through art required not just a creative transformation of individual life, but a creative transformation of the life of the city.

Understanding the experience of truth in the modern city, however, requires addressing the ways in which the authority to speak the truth is performed and recognized. A fundamental aspect of an art of living is the relationship that is stylized between life and authority. Chapter 3 addresses the ways in which the perceived 'crisis of authority' that followed the tribulations of the Paris Commune precipitated new experiments with locating legitimate authority in the dynamic energies of life and, in particular, life's expression in embodied experience. The chapter sketches a conceptual diagram of three competing models of social and cultural authority in the modern city. First, 'transcendent' authority referred back to religion, tradition, and pre-modern ideals of rural community. Second, 'experimental' authority located authority in scientific representation, drawing its legitimacy from the (supposedly) impersonal knowledge generated through scientific expertise. This aimed to replace unaccountable personal authority with the wholly objective authority of science. Finally, 'experiential' authority aimed to find a principle of legitimate authority in the vital energies of embodied experience. It is this third form of authority, and its mobilization by the arts, that will be the focus of this book. The artistic and political avant-gardes that found their home in Montmartre, I argue, used their experiments with the limits of experience in order to lend a new kind of authority to their attempts to speak truth to power. Through a reading of Baudelaire, the chapter argues that whilst experiential authority was deployed to support the biopolitical discourses of hegemonic power, it could also be used to contest dominant regimes of power. In Baudelaire's city poetry, for example, we can discern an attempt to use his lived experience of suffering to lend authority and weight to his protest against the modern city.

Chapter 4 extends this analysis to the urban culture of fin-de-siècle Montmartre, exploring the emergence of the Montmartre cultural industries following the suppression of the Commune. Montmartre emerged as a privileged space of liberal government in fin-de-siècle Paris, being allowed unprecedented cultural freedom to critique existing social values (at a safe distance from the bourgeois heart of Paris). In particular, the chapter focuses on the ways in which humour started to emerge as a particularly powerful device for contesting traditional and scientific authority. Through humorous 'experimental embodiments' of biological maladies, much Montmartre humour aimed to make the limits of life visible and felt, and hence to find ways of testing the limits of life. A key part of the attraction of the humour of Monmartre's cabarets was their ability to use performance and embodied experience to test and transgress established truths concerning the life of the city.

The book then moves on, in Part II, to take up a detailed case study of the Chat Noir 'cabaret artistique'. Three chapters in turn explore the ways in which performances at the Chat Noir worked upon the experiential life of the city at the levels of affect, representation, and perception. Chapter 5 examines the ways in which the artists and writers of the Chat Noir used humour as a way of re-imagining the possibilities of urban place and intervening in the affective economy of the city. Through ironic humour and pantomime buffoonery, which offered means of organizing contradictory semantic meanings and exploring corporeal, affective ways of inhabiting space, they attempted to negotiate the contradictions of modernity and create an experience of place that was dynamic and full of vitality without obscuring modernity's arbitrary violence, gross brutality and accelerating inequalities.

Chapter 6 addresses the representational life of the city through a study of the ethos of 'counter-display' that was discernible in the textual performances of the Chat Noir. In its mock museum catalogues and its travelogues of the city of Paris, satirizing the posture of the colonial explorer, the writers of the Chat Noir dramatized the failure of hegemonic modes of representation to establish the total orderings of life, time and history that they claimed to achieve. The Chat Noir's ethos of counter-display, I argue, aimed to mobilize a different experience of life, not as something stable and ordered, but as a form of error, an interruption of the given ordering of visibilities and representations. Through humorous counter-display, a novel form of urban vitality could emerge, a vitality that existed outside life as it was specified within the power relations of modern urban display. This life was to make itself discernible, not through the authority of representation, but through the authority of embodied experience.

The focus of Chapter 7 is upon the relationship between the life of the city and the perceptual life of the body. Through spectral, haunted landscapes of the modern city, the poets of the Chat Noir attempted to make perceptible forms of ghostly life that were invisible to the organic senses. Similarly, through the shadow theatre that made the cabaret a sensation, Montmartre writers and artists performed a derangement of the senses, a dissolution of the organic body and its supposed harmony with the urban ecology. Whilst these performances bore similarities with Wagnerian forms of total theatre, the aim was not to synthesize grand artworks, but to unwork the senses – to free them from their link to the sterile life of the city. Through the sensation of non-perceptual life, the authority of lived experience could be used to voice truths about the possibilities for new forms of urban vitality.

As Montmartre's cultural success quickly led its cabarets to lose their aura of authentic creativity and protest, the anarchist movement, with which it had been closely aligned, began to become disillusioned. Part III shifts emphasis from artistic experiments with the life of city to the political experiments of fin-de-siècle anarchism, which also found its spiritual home in Montmartre. During the 1890s, a wave of anarchist attacks connected with the violent doctrine of 'propaganda by the deed' swept across Europe. Chapter 8 discusses a little-discussed aspect of this phenomenon – the close affinities between violence and humour in the anarchist movement. The curious intersection of violence and laughter in the anarchist movement, I argue, must be interpreted in relation to the affective structure that was seen to be common to both. Violence, like humour, offered a way of disrupting the established order of visibilities and invisibilities, planting a newly materialized truth within the fabric of the city itself. Anarchism attempted to lend authority to its militant political truths through acts of violence that demonstrated extreme acts of courage. Neither anarchist humour nor anarchist violence, however, could pose a genuine challenge to hegemonic biopolitical discourses, since both could be reversed and turned back on the anarchists. By participating in the biopolitical discourses they contested, anarchism was led into a series of contradictions that were impossible to resolve.

# Acknowledgements

The research for this book was carried out in wonderfully supportive intellectual environments at Bristol, Newcastle, Warwick and Plymouth Universities. J.-D. Dewsbury has been an inspiring source of ideas, critique and friendship. Robert Mayhew offered an enormously helpful critical reading of a draft of the manuscript. Sara Jane Bailes offered valuable advice and support in the project's early stages. Kevin Hetherington and Mark Jackson also contributed invaluable readings of early drafts of the manuscript. Many of the theoretical arguments elaborated in the book were sparked in the non-representational geographies reading group at Bristol. They were further developed in an incredibly productive series of theory retreats with the Authority Research Network. Subsequent ideas and drafts have been developed with the support of numerous friends and colleagues, as well as at several conferences.

Staff at the Archives de la Préfecture de Police de Paris, the Archives Nationales, the Bibliothèque Sainte-Geneviève, Bristol University Library, The British Library, the Musée d'Orsay, the Musée de Montmartre, Plymouth University library, and the V&A library have been incredibly helpful.

Val Rose, Margaret Younger, Emily Ruskell and the rest of the team at Ashgate have been extremely helpful and supportive throughout the publishing process.

Hugh and Anthea Brigstocke, as well as Sophie, Marcus, Freddie and Claudia Green, offered unfailing support, encouragement and good company throughout the years that it took to complete this book. David and Lynda Blencowe shared an idyllic seaside refuge at times of stress and crisis. Carol and Danny Doran's hospitality and generosity transformed a potentially gruelling few months of writing into an enormously enjoyable time in which I felt welcomed into a second family.

Finally, the book could never have been completed without Claire Blencowe's unwavering love, inspiration and friendship.

The research for this book was generously funded by the UK Economic and Social Research Council. Chapter 5 was originally published as 'Defiant Laughter: Humour and the Aesthetics of Place in Late 19th century Montmartre', *Cultural Geographies* 19 (2012), pp. 217–35. It is reproduced here with permission from Sage. Some fragmentary material throughout the book was also published in 'Immanent Authority and the Performance of Community in Late Nineteenth Century Montmartre', *Journal of Political Power* 6 (2013), pp. 107–26. This material is reproduced here with permission from Taylor and Francis.

# Chapter 1

# Introduction:
# Art, Life, and the Experience of Modernity

## Fin-de-siècle Montmartre

In a withering attack on the 'end of the notables' that he believed to have accompanied the beginnings of the Third Republic of France, the conservative historian Daniel Halévy picked out one symptom of the withering of true French spirit and vitality. This was the explosion of carnivalesque popular culture in the cafes and cabarets of Montmartre. In *Pays Parisiens,* Halévy names a single event as the focus of this decadent spirit of modernity: the riotous opening of the Chat Noir (Black Cat) cabaret on 18 November 1881 at 84 Boulevard Rochechouart, at the foot of the hill of Montmartre.[1] The Chat Noir quickly became a symbolic home for a much wider series of cultural experiments that paradoxically made Montmartre, a self-professed space of marginality and exclusion, a focal point of fin-de-siècle urban modernity.

Montmartre has long enjoyed an almost mythical place in the French imagination. It was on this druidic holy hill near the Roman city of Lutetia that a bishop known as Dionysius was martyred during a persecution of Christians ordered by the Emperor Decius, shortly after the year 250. Having refused to renounce his faith after long and grizzly tortures, Dionysius was finally beheaded. The moment his severed head struck the earth, however, Dionysius calmly reached down and plucked it off the floor, accompanied by a host of angels singing God's praises. Dionysius then walked for ten kilometres with his head in his arms, preaching sermons along the way, before indicating his preferred resting place at the site of what later became the Basilica de Saint-Denis. Saint Denis became a patron saint of France, and the Basilica was to become the standard burial place of the French kings until the eighteenth century.[2] The name 'Montmartre' is said to derive from 'Mons Martyrum' (Martyr's Mount), in reference to this episode.

From early in the history of France, then, Montmartre was a symbolic heart of religious and royal authority. Over the second half of the nineteenth century, however, it started to cast off this symbolism. In 1860 the village had been annexed into a rapidly expanding Paris, soon losing its village-like character to become a

---

1   On the end of the notables, see Daniel Halévy, *La Fin Des Notables* (Paris, 1930). On Montmartre, see Daniel Halévy, *Pays Parisiens: Portrait De La France* (Paris, 1929).

2   Gabrielle Spiegel, 'The Cult of Saint Denis and Capetian Kingship', *Journal of Medieval History* 1(1) (1975).

poor – and thus potentially dangerous – working-class neighbourhood.[3] It retained a distinctive and picturesque rural character, however, thanks largely to its windmills, vineyards, and steep winding streets, which attracted painters such as Camille Corot and Camille Pisarro in the middle of the century (Figure 1.1). After the renovations of Paris during the Second Empire, during which much of the city of Paris was demolished and rebuilt according to a highly rationalized, geometrical, monumental plan, the winding lanes of Montmartre, which had not undergone this transformation, became a reminder of what had been lost in the process of modernization.[4]

Along with its reputation for rural tranquillity, however, Montmartre also had a long history of disorder and unruliness. The boulevard tracing the lower edge of Montmartre followed the line of the old Mur des Fermiers Généraux, a tax wall where a duty had to be paid on goods passing into the city of Paris. The area beyond the wall, in present-day Montmartre, had long been a place of popular entertainment, since alcohol could be enjoyed more cheaply there than within the city's boundaries. Long after the wall was demolished, the area retained something of this popular spirit. It remained poor, freedom-loving, and populated with petty criminals, procurers and prostitutes. And in 1871 it became the symbolic home of the urban revolution known as the Paris Commune, during which workers seized control of the city of Paris and declared its autonomy from the rest of France. This led Montmartre to become associated with the spirit of revolt, revolution, and rejection of the authority of the new Republic.[5] Montmartre was henceforth to be associated with a general crisis of authority in the early Third Republic.

Whilst aspects of the story of Montmartre have been told many times, much of the writing on fin-de-siècle Montmartre has remained descriptive and anecdotal.[6] In recent years, however, researchers have contributed to a more social scientifically nuanced approach to the history of the neighbourhood.[7]

---

3   Louis Chevalier, *Labouring Classes and Dangerous Classes in Paris During the First Half of the Nineteenth Century*, trans. Frank Jellinek (London, 1973).

4   David Harvey, *Paris, Capital of Modernity* (New York, 2003), François Loyer, *Paris Nineteenth Century: Architecture and Urbanism* (New York, 1988), David Pinkney, *Napoleon III and the Rebuilding of Paris* (Princeton, NJ, 1958).

5   On the Commune, see M. Castells, 'Cities and Revolution: The Commune of Paris, 1871', *The City and the Grassroots: A Cross-Cultural Theory of Urban Social Movements* (Berkeley, 1983), Henri Lefebvre, *La Proclamation De La Commune: 26 Mars 1871* (Paris, 1965), Robert Tombs, *The Paris Commune, 1871* (London, 1999).

6   For example, Sylvie Buisson and Christian Parisot, *Paris Montmartre: Les Artistes and Les Lieux, 1860–1920* (Paris, 1996), Philippe Jullian, *Montmartre*, trans. Anne Carter (Oxford, 1977), Mariel Oberthur, *Le Cabaret Du Chat Noir À Montmartre* (Geneva, 2007), Charles Rearick, *Pleasures of the Belle Epoque: Entertainment & Festivity in Turn of the Century France* (New Haven, 1985).

7   Maurice Agulhon, 'Paris: A Traversal from East to West', in Pierre Nora (ed.), *Realms of Memory: The Construction of the French Past, Iii, Symbols* (New York, 1998), Nicholas Hewitt, 'Shifting Cultural Centres in Twentieth-Century Paris', in M. Sheringham (ed.), *Parisian Fields* (London, 1996), Nicholas Hewitt, 'From "Lieu De Plaisir" to "Lieu De

**Figure 1.1     The hill of Montmartre, seen from Place Saint-Denis. Engraving, 1866**

Cultural historians, in addition, have observed the extent to which Montmartre's cabaret culture anticipated the later movements of the historical avant-garde. Phillip Dennis Cate and Mary Shaw, in particular, have uncovered a fascinating collection of art, music and literature that signals a clear rejection of the established principles of art, and whose implicit challenge to the very institution marks it with a distinctively avant-gardist spirit.[8] What they largely neglect, however, is the social and urban environment in which these creative experiments were conducted. Research into the artistic culture of Montmartre has generally focused upon specific individuals. As Gabriel Weisberg observes, 'the examination of these individuals has been heavily reportorial and often devoid of the larger context of the very culture of Montmartre – of the pervading social environment that encouraged and rewarded experimentation'.[9] Yet the urban

Mémoire": Montmartre and Parisian Cultural Topography', *French Studies* LIV(4) (2000), Jeffrey Jackson, 'Artistic Community and Urban Development in 1920s Montmartre', *French Politics, Culture & Society* 24(2) (2006), N. Kenny, 'Je Cherche Fortune: Identity, Counterculture, and Profit in Fin-De-Siècle Montmartre', *Urban History Review* 32 (2004).

8     Phillip Dennis Cate and Mary Shaw (eds), *The Spirit of Montmartre: Cabarets, Humor, and the Avant-Garde, 1875–1905* (New Brunswick, NJ, 1996).

9     Gabriel Weisberg, 'Montmartre's Lure: An Impact on Mass Culture', in Gabriel Weisberg (ed.), *Montmartre and the Making of Mass Culture* (New Brunswick, NJ, 2001), p. 2.

environment, I will argue, is inseparable from these experiments: it was both the subject and the medium of many of the most important aspects of fin-de-siècle Montmartre culture.

The collection of essays in Weisberg's *Montmartre and Mass Culture* goes some way to addressing this, highlighting, for example, the anarchist political culture of Montmartre; the place of women and religion in Montmartre life; and popular concern for the physical health of Montmartre children and the population at large.[10] In addition, the exhibition catalogue *Toulouse-Lautrec and Montmartre* offers portraits of the social environment in which Henri de Toulouse-Lautrec – by far the most well-known painter of everyday Montmartre life – lived and worked.[11] An account of the longer history of Montmartre, finally, is offered by the historian Louis Chevalier, who traces the development of Montmartre society and culture from the early nineteenth century to the Second World War. Chevalier describes the Montmartre of the early 1880s as 'a festival of spirit'.[12] The opening of the Chat Noir cabaret, he notes, marked a shift from 'calm conversation about art and literature' to 'an unquenchable verve and a knack for improvization. An eloquence of buffoonery and incoherence'.[13] Chevalier describes the unique character of Montmartre during this time as resulting from an unusual symbiotic relationship between three different classes of people: impoverished artists; wealthy bourgeois; and an underworld of pimps, prostitutes, vagrants and criminals.

The culture of late nineteenth century Montmartre has been interpreted as anticipating both the avant-gardes of the early twentieth century, and also modern mass culture.[14] On the one hand, it witnessed a host of innovative experiments with destabilizing conventional aesthetic codes and sensibilities, and questioning the very possibility or desirability of art in modernity. On the other hand, its rapid successes resulted in the establishment of financially lucrative venues such as the Moulin Rouge (Red Mill) dancehall, and contributed to its growing reputation as a space, not of artistic creativity and originality, but of hedonism, decadence,

---

10    Gabriel Weisberg (ed.), *Montmartre and the Making of Mass Culture* (New Brunswick, NJ, 2001).

11    Richard Thomson, Phillip Dennis Cate and Mary Weaver Chapin (eds), *Toulouse Lautrec and Montmartre* (Washington, 2005).

12    Louis Chevalier, *Montmartre du Plaisir et du Crime* (Paris, 1995), p. 152.

13    Ibid. 'Une fête d'esprit ... Si les artistes aimaient se retrouver et discuter avec calme et littérature à la Grande Pinte, il n'en fut pas de même au Chat Noir ... [Salis] y présenta lui-même ses camarades avec une verve inlassable et des dons prodigieux d'improvisateur'. See also Anne de Bercy and Armand Ziwès, À Montmartre ... Le Soir. Cabarets *et Chansonniers D›Hier* (Paris, 1951).

14    Cate and Shaw (eds), *The Spirit of Montmartre: Cabarets, Humor, and the Avant-Garde, 1875–1905*, Bernard Gendron, *Between Montmartre and the Mudd Club: Popular Music and the Avant-Garde* (Chicago, 2002), Weisberg (ed.), *Montmartre and the Making of Mass Culture*, Steven Moore Whiting, *Satie the Bohemian: From Cabaret to Concert Hall* (Oxford, 1999).

pleasure – and profit.[15] The multi-sensual shadow plays or 'ombres chinoises' of the Chat Noir, in addition, anticipated the development of the cinematic moving image, and became a draw to audiences across France (see Chapter 6). But accounts of Montmartre culture have either failed to place it within its wider social, economic and geographical context, or else have framed it within well-worn debates concerning the relation between art and the market. The scholarly debate often boils down to an argument concerning whether or not Montmartre bohemia contributed to an anti-capitalist, genuinely avant-gardist challenge to modern art and life, or whether instead it marked art's final submission to market forces, and the last betrayal of the principles of the Paris Commune.

One striking omission from these accounts is a clear analysis of the ways in which Montmartre popular and avant-garde culture engaged with its urban environment. No studies, for example, have discussed the emergence of the Montmartre creative milieu in relation to the powerful new 'biopolitical' discourses, which dominated French urban government during this time, relating to the health, vitality and life of the modern city. There is little understanding of how Montmartre writers and artists depicted and conceptualized the urban environment, either of Montmartre itself, or the rest of the city from which Montmartre fiercely disassociated itself. What analysis does exist makes use of simplistic motifs of marginality and transgression, and fails to offer much insight into the ways in which Montmartre artists' experiments with urban community meshed with emerging biopolitical strategies of urban governance that emphasized the pluralization of individual freedom, the impossibility and undesirability of total government control, and a vigorous concern with the health, vitality and dynamism of the population and its urban environment.[16]

In a new Republic with a love of individual liberty, Montmartre quickly emerged as a semi-mythical urban area in which exciting new freedoms were seen to be freely at hand. In order to understand the ways in which these freedoms emerged and were transformed, however, it is necessary to look at Montmartre culture, not just in terms of its place in the flows of capital, but also in terms of its interventions in an economy of life, lived experience and urban vitality. That is to say, it is necessary to ask how the creative community in Montmartre attempted to alter the urban environment, intervening in the life, vitality and experience of the city, and thereby contributing to experiments with novel arts of urban life. In order better to understand the political and communal aspirations of the Montmartre community, in other words, we need to situate the culture of Montmartre, not within an economy of representation which places it 'inside' or 'outside' dominant frameworks of power, but within an economy of life and lived experience.

---

15   Rearick, *Pleasures of the Belle Epoque: Entertainment & Festivity in Turn of the Century France*.

16   On Montmartre as a space of transgression, see, for example, Richard Sonn, 'Marginality and Transgression: Anarchy's Subversive Allure', in Gabriel Weisberg (ed.), *Montmartre and the Making of Mass Culture* (New Brunswick, NJ, 2001).

The question of the life of the city, and the authority by which it could be known and contested, I will argue in this book, was at the very heart of the creative endeavours of Montmartre writers and artists. Until the present study, however, this important dimension of the historical geography of the modern metropolis has remained largely ignored.

## The Chat Noir

This book analyses the urban culture of several cafes, cabarets and other performance spaces in Montmartre. The most well-known performance space – and arguably the most important – was the Chat Noir cabaret. The dramatic growth of the Montmartre cultural industries – which many advocates of contemporary arts-led urban regeneration have attempted to replicate via the manufacture of supposedly 'neo-bohemian' urban spaces – is often attributed to this single venue. The Chat Noir was not a conventional cabaret (a proletarian drinking venue), but a 'cabaret artistique', a novel kind of café where artists would come to perform and display their works, as well as exchange ideas, in a relaxed and convivial environment.[17] Curious publics could also come to watch, and the eclecticism of the cabaret form soon proved to be enormously popular. The first Chat Noir building was tiny, comprising only two rooms. By 1885, however, its success enabled it to move to larger premises, with three floors, at 12 Rue de Laval (now rue Victor Massé). Before long, imitations started cropping up, first around Montmartre, and later, as it became more and more famous, in the provinces and finally as far away as Barcelona, St Petersburg and Kraków.

In fact, however, the Chat Noir was not the first 'cabaret artistique'. The Grande Pinte opened in 1878, and other venues such as the Rat Mort, the Café Guerbois, the Bon Bock, and the Nouvelle-Athènes were also popular gathering spots, attracting Impressionists and other artists and intellectuals of their generation.[18] What the Chat Noir added to the mix, however, was its sending up of established aesthetic and literary values and debates, and its dissemination of a spirit of anarchic humour, parody and satire – what was often called 'fumisme'.[19]

---

17    See Lisa Appignanesi, *The Cabaret* (London, 1984), Mary Weaver Chapin, 'The Chat Noir & the Cabarets', in Richard Thomson, Phillip Dennis Cate, and Mary Weaver Chapin (eds), *Toulouse Lautrec and Montmartre* (Washington, 2005), Oberthur, *Le Cabaret Du Chat Noir À Montmartre*, Harold Segel, *Turn-of-the-Century Cabaret: Paris, Barcelona, Berlin, Munich, Vienna, Cracow, Moscow, St. Petersburg, Zürich* (New York, 1987).

18    Phillip Dennis Cate, 'The Social Menagerie of Toulouse-Lautrec's Montmartre', in Richard Thomson, Phillip Dennis Cate and Mary Weaver Chapin (eds), *Toulouse Lautrec and Montmartre* (Washington, 2005).

19    Phillip Dennis Cate, 'The Spirit of Montmartre', in Phillip Dennis Cate and Mary Shaw (eds), *The Spirit of Montmartre: Cabarets, Humor and the Avant-Garde, 1875–1905* (New Brunswick, 1996), Janet Whitmore, 'Absurdist Humor in Bohemia', in Gabriel Weisberg (ed.), *Montmartre and the Making of Mass Culture* (New Brunswick, NJ, 2001).

In this respect, the Chat Noir cultivated an anarchic spirit that was to form important cross-fertilizations with the anarchist movement, not to mention being echoed by later avant-garde movements such as Dada. The adjective 'chatnoiresque' was coined to describe the cabaret's unique blend of humour, irreverence, political critique, and artistic creativity.

The cabaret was the result of a meeting between a former art student called Rodolphe Salis, and a talented poet called Émile Goudeau. Salis, the son of a brewer, had moved from Switzerland to Paris with ambitions to be an artist, and had spent some time studying at the Ecole des Beaux-Arts before being rejected at the official Paris Salon.[20] Taking an entrepreneurial turn, with the financial backing of his father, he decided to open a small café at the foot of the butte Montmartre. The name he chose, the Chat Noir, evoked the literature of Charles Baudelaire and Edgar Allen Poe, the paintings of Édouard Manet, classic French folktales, and the supernatural world of witches and magic (see Chapter 2). The term 'Chat Noir' was also a sexual double entendre. This combination of art, literature, mystery and sexual suggestion suited Salis' aims perfectly, and the black cat, with all its innuendo and mystery, was to become an enduring symbol of Montmartre.

By chance, Salis met the leader of the Hydropathes group of bohemian writers and artists, Émile Goudeau, at the Grande Pinte cabaret in November 1881. The Hydropathes were no longer welcome at their former meeting venue in the Latin Quarter, and Salis and Goudeau quickly came to an arrangement that the artists would come to the Chat Noir to meet, drink, and share work and ideas. The success of the Chat Noir arose out of a combination of Goudeau's artistic talents and Salis' extraordinary talent for generating publicity and controversy. Its first successes came from word of mouth. Soon, taking advantage of the Republic's newly liberalized press laws, Salis and Goudeau set up a literary journal, *Le Chat Noir*, which was distributed across Paris. The journal, edited by Goudeau, was at once an outlet for the group's artistic and literary ambitions, and also a means of promoting the venue across Paris. One advert in the journal proclaimed, with characteristic hyperbole:

> The Chat Noir is the most extraordinary cabaret in the world. You mingle with the most famous men in Paris, who meet there with foreigners from every corner of the globe. [Novelists] Victor Hugo, Émile Zola, Barbey d'Aurevilly, the inseparable [zoologist] Mr Brisson, and the austere [politician] Gambetta chat informally with [art critic] Gaston Vassy and [banker and collector] Gustave Rothschild. People hurry in and press themselves inside. It's the greatest success of the age! Come on in! Come on in![21]

---

20   Georges Auriol, 'Rodolphe Salis Et Les Deux "Chat Noir"', *Mercure de France* 9 (1926).

21   Anon., 'Le Chat Noir', *Le Chat Noir*, 8 April 1881. 'LE CHAT NOIR Est le cabaret le plus extraordinaire du monde. On y coudoie les hommes les plus illustres de Paris, qui s'y rencontrent avec des étrangers venue de tous les points du globe.

The paper was an important publicity organ, establishing the humorous tone and spirit of the cabaret through the irreverent style of the articles and the illustrations by Montmartre artists such as Théophile Steinlen. The cabaret soon became a favourite destination for writers, poets, musicians, and artists, who were both the clients and the entertainment, since they lost no opportunity to donate pictures, recite poetry, and play music. Salis managed to obtain permission to put a piano in the cabaret (something that was usually banned), and thus started the tradition of popular song which is now strongly associated with Montmartre.[22]

Whilst the cabaret is now a famous site of Parisian modernity, the specifically urban and biopolitical dimensions of the performances at the Chat Noir remain unexplored. This book contributes to current understandings of the Chat Noir by conducting a detailed analysis of the spatial elements of the performances, parades, and literature there. Through analyses of its imaginative geographies of Paris, its parodies of museum spaces, and the synaesthetic performances that aimed to creatively work upon the geographies of the body, I will draw out the ways in which it tested and challenged bourgeois assumptions concerning the experiential life of the city.

## Modernity and the Genealogy of Arts of Living

Outside the Chat Noir cabaret, a sign could be seen exhorting: 'Passer-by, Halt! Be modern!'[23] But what did it mean to be modern in the spaces of Parisian counter-culture? The term 'modernity', with all its ambivalence and polysemy, is at once suggestive and disturbing. It troubles with its evocation of alienating forces of rationalization and homogenization: the imposition of geometric grids; the sterilization of urban environments; the violent elimination of ambiguity. Yet something about it retains a certain allure: the technological production of unimaginable new powers and capacities; the thrill of the new and the unexpected; the production of dramatically new forms of experience and subjectivity. Implicit in each of these responses, perhaps, is a certain attitude towards creativity and difference. Some aspects of modernity seem violently to close down creativity, but others provide the means for an enormous proliferation of new forms of creativity and difference. In modernity, the potential for stylizing

Victor Hugo, Émile Zola, Barbey d'Aurevilly, l'inséparable M. Brisson, l'austère Gambetta s'y tutoient avec MM. Gaston Vassy et Gustave Rothschild, on s'y foule, on s'y presse.
C'EST LE PLUS GRAND SUCCÈS DE L'ÉPOQUE
ENTREZ! ENTREZ !'
    22    Bercy and Ziwès, À Montmartre ... Le Soir. Cabarets et Chansonniers D'Hier, Horace Valbel, *Les Chansonniers et les Cabarets Artistiques* (Paris, 1895), Pierre Veber, 'Les Cabarets Artistiques et la Chanson', *La Revue D'Art Dramatique*, 15 December 1889.
    23    'Passant, Arrête-toi! Sois moderne!

new arts of life – novel ways of creating subjectivity and experience – can appear almost unbounded.[24]

In the twenty-first century, however, with the consolidation of neoliberal modes of government, the avant-garde ideal of living life as a work of art has become, not so much a utopian aspiration, as a daunting necessity. As Zygmunt Bauman puts it:

> Our lives, whether we know it or not and whether we relish the news or bewail it, are works of art. To live our lives as the art of life demands, we must, just like the artists of any art, set ourselves challenges which are … well beyond our reach, and standards of excellence that vexingly seem to stay stubbornly far above our ability … We need to *attempt the impossible*.[25]

In these words Bauman captures something of the anxiety and ambiguity of the experience of the contemporary self – the way in which the neoliberal subject has become trapped by her own dreams of freedom, self-invention, and rejection of fixed identities. Via neoliberal techniques of government, the opportunity to stylize new arts of living, following the decline in transcendent authority and universal moral codes, has been seized for the production of entrepreneurial selves, forms of 'human capital' who are responsible for managing and reinventing themselves in order to compete more effectively in the market.[26] Disoriented by a creeping recognition that the source of her domination is her own freedom, making her at once master and slave of herself, the neoliberal subject is left grasping at the impossible, still intoxicated by the possibility of stylizing her life as a work of art, but increasingly fearful of the consequences of failing to do so in line with the demands of the market.

For Michel Foucault, one strategy for undermining the modern injunction that we all become creatively self-fashioning individuals, adopting an economic rationality of self-as-enterprise, is to construct a genealogy of 'arts of living' or 'aesthetics of existence' that at once traces the emergence of such neoliberal discourses and also reveals the alternatives that were defeated or pre-empted. Foucault's late turn to the genealogy of ethics involved an attempt to rethink the possibilities for 'arts of living' that distance themselves from the entrepreneurial rationalities of neoliberalism. 'What strikes me', Foucault remarks, 'is the fact that, in our society, art has become something that is related only to objects

---

24   On spaces of modernity in Europe, see for example Richard Dennis, *Cities in Modernity: Representations and Productions of Metropolitan Space, 1840–1930* (Cambridge, 2008), David Gilbert, David Matless and Brian Short (eds) *Geographies of British Modernity* (Oxford, 2003), Kevin Hetherington, *The Badlands of Modernity: Heterotopia and Social Ordering* (London, 2002), Miles Ogborn, *Spaces of Modernity: London's Geographies, 1680–1780* (New York, 1998).

25   Zygmunt Bauman, *The Art of Life* (Cambridge, 2008), p. 20, emphasis in original.

26   Maurizio Lazzarato, 'Neoliberalism in Action: Inequality, Insecurity and the Reconstitution of the Social', *Theory, Culture & Society* 26(6) (2009).

and not to individuals or to life. That art is something which is specialized or done by experts who are artists. But couldn't everyone's life become a work of art?'[27] Foucault's final books and lectures explored the potential for reactivating such arts of living in modernity through an examination of Greek, Roman and early Christian aesthetics of existence.[28] The history of arts of living, Foucault comments, has been largely neglected since Jacob Burckhardt's classic account of Renaissance Italy, with the important exception of Walter Benjamin's excavation of Charles Baudelaire's writings on modernity and Stephan Greenblatt's work on Renaissance self-fashioning.[29]

Extending this Foucauldian approach, in this book I revisit an episode in the history of urban arts of living that has acquired a prominent place in neoliberal discourses of entrepreneurial self-invention. The rise of a bohemian avant-garde in Montmartre, quickly followed by a parasitical culture industry which converted the impoverished neighbourhood into a thriving centre of alternative modern urban culture, is a story that has frequently been cited in narratives of culture-led urban regeneration and the neo-bohemian 'creative city'.[30] As Michel Foucault has analysed, however, the contemporary hegemony of neoliberal logics of creative self-fashioning and entrepreneurship can be countered through forms of thinking that trace the emergence of such discourses, and in doing so, experiment with the possibilities for moving beyond them. Foucault's tentative experiments with a history of arts of living should be viewed as a direct response to the emerging forms of neoliberal subjectivity that he diagnosed so acutely in *The Birth of*

---

27  Michel Foucault, 'On the Genealogy of Ethics: An Overview of a Work in Progress', in Paul Rabinow (ed.), *Ethics, Subjectivity and Truth: Essential Works of Foucault 1954–1984, Volume One* (London, 2000), p. 261.

28  Michel Foucault, *The Care of the Self: The History of Sexuality, Volume Three*, trans. R. Hurley (London, 1990), Michel Foucault, *The Use of Pleasure: The History of Sexuality, Volume Two*, trans. R. Hurley (London, 1992), Michel Foucault, *Fearless Speech* (2001), Michel Foucault, *The Hermeneutics of the Subject: Lectures at the Collège de France 1981–1982* trans. Graham Burchell (Basingstoke & New York, 2005), Michel Foucault, *The Government of Self and Others: Lectures at the Collège de France, 1982–1983* (Basingstoke, 2010), Michel Foucault, *The Courage of Truth: The Government of Self and Others, Volume Two. Lectures at the Collège de France, 1983–1984* (Basingstoke, 2011).

29  Foucault, *The Use of Pleasure: The History of Sexuality, Volume Two*, p. 11. See Jacob Burckhardt, *The Civilisation of the Period of the Renaissance in Italy*, trans. S. Middlemore (London, 1878), Stephen Greenblatt, *Renaissance Self-Fashioning: From More to Shakespeare*, 2nd ed. (Chicago & London, 1980).

30  Richard Florida, *The Rise of the Creative Class: And How It's Transforming Work, Leisure, Community and Everyday Life* (New York, 2002), Richard Florida, *Cities and the Creative Class* (New York & Abingdon, 2005), John Hannigan, 'A Neo-Bohemian Rhapsody: Cultural Vibrancy and Controlled Edge as Urban Development Tools in the "New Creative Economy"', in Tim Gibson and Mark Lowes (eds), *Urban Communication: Production, Text, Context* (Lanham, MD, 2006), Richard Lloyd, *Neo-Bohemia: Art and Commerce in the Postindustrial City* (London & New York, 2005).

*Biopolitics.*[31] 'Foucault's own insistence in thinking about the subject constituted as practices works both *with* and *against* neoliberal subjectivity and neoliberal conceptions of freedom, truth, and reality'.[32] Foucault's genealogies of ethics, that is, aimed to come to terms with the potentials for arts of living that are not reducible to the entrepreneurial rationality of 'homo economicus'. In the same spirit, the present study will document the 'arts of living' in late nineteenth-century Montmartre both in order to chart the emergence of an urban discourse of creativity that has been very powerful in contemporary urban regeneration discourses, and to test the limits of this discourse to uncover points of weakness and fracture.

Foucault's account of the art of life has frequently been subjected to accusations of a drearily conservative 'return to the subject'– a lapse, perhaps, into a geriatric existentialism.[33] This is somewhat unfair, since Foucault's main concern was to point to the relationships between the government of self and the government of others, with a view to the invention of new ways of speaking truth to power.[34] Nevertheless, his two published books (as opposed to his lecture series) did focus on techniques of individual self-mastery. By contrast, my focus in what follows is not so much on creative interventions on individual life, but on the ways in which Montmartre artists and political activists attempted to creatively intervene in the *life of the city*: the collective ecological environment that helped condition the conditions of possibility for life, experience and knowledge in modernity.

## Art, Life and Spectacle

Foucault's theorization of the relationship between art and life can be contrasted with a type of Marxist perspective that has been highly influential in scholarly literature on urban avant-garde movements of the early and mid-twentieth century. The 'avant-garde' has often been defined in terms of the determination to overcome the separation between artistic creativity and the spaces of everyday life.[35] Theories of the avant-garde frequently incorporate a spatial sensibility into

---

31    Michel Foucault, *The Birth of Biopolitics: Lectures at the Collège de France, 1978–1979*, ed. Michel Sellenart, trans. Graham Burchell (Basingstoke & New York, 2008).

32    Andrew Dilts, 'From "Entrepreneur of the Self" to "Care of the Self": Neoliberal Governmentality and Foucault's Ethics', *Foucault Studies* 12 (2011), p. 132.

33    See for example Peter Dews, 'The Return of the Subject in Late Foucault', *Radical Philosophy* 51 (1989), Terry Eagleton, *The Ideology of the Aesthetic* (Oxford, 1990), Richard Wolin, 'Foucault's Aesthetic Decisionism', *Telos* 67 (1986).

34    Foucault, *The Government of Self and Others: Lectures at the Collège de France, 1982–1983*.

35    See for example Peter Bürger, *Theory of the Avant-Garde*, trans. Michael Shaw (Manchester, 1984), Matei Călinescu, *Five Faces of Modernity: Modernism, Avant-Garde, Decadence, Kitsch, Postmodernism* (Durham, NC, 1987), Richard Murphy, *Theorizing*

theories of the links between art and life.[36] In urban theory, the most powerful statement of such a project has come from theorists associated with the Situationist avant-garde of the 1960s such as Guy Debord and Henri Lefebvre, who suggested in *The Production of Space* that,

> On the horizon, then, at the further edge of the possible, it is a matter of producing the space of the human species – the collective (generic) work of the species – on the model of what used to be called 'art'; indeed, it is still so called, but art no longer has any meaning at the level of an 'object' isolated by and for the individual.[37]

One key emphasis in Debord's critique of the 'society of the spectacle' and Lefebvre's critique of everyday life was upon the changing geographies of the city.[38] Debord critiqued the ways in which the capitalist spectacle had succeeded in completely colonizing the urban environment. Although the word 'spectacle' is often used loosely in connection with specific sites – trade fairs, exhibitions, department stores, and so on – Debord used the concept of 'spectacle' to refer to a total occupation of urban space. In Debord's theory, '[t]he spectacle dominated social life and space, homogenizing and fracturing space, unifying and separating, and becoming "the perfection of separation *within* human beings"'.[39] This theory

---

*the Avant-Garde: Modernism, Expressionism, and the Problem of Postmodernity* (Cambridge, 1999), Peter Osborne, *The Politics of Time: Modernity and the Avant-Garde* (London & New York, 1995), R. Poggioli, *Theory of the Avant-Garde*, trans. G. Fitzgerald (1968).

36   On spatialities of the avant-garde, see, for example, Alastair Bonnett, 'Situationism, Geography, and Poststructuralism', *Environment and Planning D: Society and Space* 7(2) (1989), Alastair Bonnett, 'Art, Ideology and Everyday Space: Subversive Tendencies from Dada to Postmodernism', *Environment and Planning D: Society and Space* 10 (1992), J.-D. Dewsbury 'Avant-Garde/Avant-Garde Geographies', in Rob Kitchen & Nigel Thrift (eds) *International Encyclopedia of Human Geography* (Amsterdam, 2009), pp. 252–6, Anja Kanngieser, *Experimental Politics and the Making of Worlds* (Farnham, 2013), David Pinder, 'Subverting Cartography: The Situationists and Maps of the City', *Environment and Planning A* 28 (1996), David Pinder, '"Old Paris Is No More": Geographies of Spectacle and Anti-Spectacle', *Antipode* 32(4) (2000), David Pinder, *Visions of the City: Utopianism, Power and Politics in Twentieth-Century Urbanism* (Edinburgh, 2005), Simon Rycroft, *Swinging City: A Cultural Geography of London 1950–1974* (Farnham, 2011), Alexander Vasudevan, 'Symptomatic Acts, Experimental Embodiments: Theatres of Scientific Protest in Interwar Germany', *Environment and Planning A* 39(8) (2007).

37   Henri Lefebvre, *The Production of Space*, trans. Donald Nicholson-Smith (Oxford, 1991), p. 422.

38   Kevin Hetherington, *Capitalism's Eye: Cultural Spaces of the Commodity* (Abingdon, 2007), Pinder, '"Old Paris Is No More": Geographies of Spectacle and Anti-Spectacle'.

39   Pinder, '"Old Paris Is No More": Geographies of Spectacle and Anti-Spectacle', p. 365, emphasis in original.

hinges upon a categorical contrast between life and representation. According to Debord, the capitalist spectacle reduces life, as a dynamic, direct, richly embodied form of energy and experience, to immobility, abstraction, and lifelessness. 'Everything that was directly lived has receded into a representation'.[40] The spectacle assimilates 'all the *fluid* aspects of human activity so as to possess them in congealed form', and urban planning contributes to this 'petrification of life'.[41] The spectacle destroys genuine life and lived experience.

A similar politics is also discernible in Lefebvre's urban theory. His critique of everyday life develops a corporeal politics committed to the creation of 'moments' of potential.[42] Moments include such varied phenomena as love, games, rest, knowledge, poetry and justice. A moment is a 'fleeting but decisive sensation', an activity in which 'a temporality of rupture and spontaneity tends towards a unification of the festival and everyday life'.[43] A radical urban politics of culture, according to Lefebvre, must aim at 'the uniting of the Moment and the everyday, of poetry and all that is prosaic in the world, in short, of Festival and ordinary life, on a higher plane than anything which has hitherto been accomplished'.[44] This politics relies upon Lefebvre's sociological adaptation of Henri Bergson's vitalist philosophy of experience.[45] Lefebvre rejects capital's reduction of qualitative temporal rhythms to a linear, homogeneous temporality, as well as the reduction of living processes to abstract quantities.[46] Progressive politics, he suggests, is focused upon interrupting the mechanical, linear rhythms of abstract space with the organic, natural rhythms of embodied, corporeal space, liberating the temporality of natural, organic life from its artificial constraints in everyday life. The avant-garde interrupts the mundane repetitions of everyday life with a burst of vital, creative life.

Debord's and Lefebvre's theories of urban life have done much to aid scholarly understandings of the spatial politics of the avant-garde. Yet this theoretical approach to the intersection of art of life has serious flaws. As Kevin Hetherington argues, one problem with the framework of the spectacle is that it effectively denies individuals' existence 'as a creative force in the world with a self-reflective consciousness and imagination'.[47] He stresses the need to engage with the ways in

---

40   Guy Debord, *Society of the Spectacle* (Detroit, 1983), p. 1.

41   Ibid., 17, 95.

42   Henri Lefebvre, *Critique of Everyday Life, Volume Two: Foundations for a Sociology of the Everyday*, trans. John Moore (London & New York, 2002).

43   Kirsten Simonsen, 'Bodies, Sensations, Space and Time: The Contribution from Henri Lefebvre', *Geografiska Annaler B* 87 (2005), p. 8.

44   Lefebvre, *Critique of Everyday Life, Volume Two: Foundations for a Sociology of the Everyday*, p. 349.

45   Henri Bergson, *Matter and Memory*, trans. W. Palmer and Nancy Paul (London, 1911).

46   Benjamin Fraser, 'Toward a Philosophy of the Urban: Henri Lefebvre's Uncomfortable Application of Bergsonism', *Environment and Planning D: Society and Space* 26(2) (2008).

47   Hetherington, *Capitalism's Eye: Cultural Spaces of the Commodity*, p. 70.

which individuals act in and engage with spaces of consumption in complex and heterogeneous ways, not simply being possessed by the commodity's witchcraft, but instead taking possession of it, and creating something new out of it. Such a perspective requires a rejection of the simplistic distinction between consumer culture, which exists as a set of activities that are wholly complicit with the dominant forms of subject-formation, and an oppositional culture which carves out temporary enclaves that are free from the power of capital.

For all its revolutionary passion, the theory of spectacle is based upon an overly Romanticist critique of urban abstraction and alienation, and a correspondingly uncritical celebration of life, embodiment, and lived experience. In Debord's and Lefebvre's urban theory of the avant-garde, creative experiments need to be devised that are capable of wresting temporary spaces of freedom from sterile, lifeless cities that are dominated by consumerism and a culture of dumb mass consumption. Genuine urban life must be liberated from the oppressive powers of capital. Yet if we take on board Foucault's excavation of the pivotal role of 'biopolitical' discourses in the constitution of modern experience, I wish to argue, it will be necessary to acknowledge the ways in which modern urbanism, far from deadening life and destroying experience, went to great lengths pains to maximize the experiential life of the city. Far from being destroyed, life became one of the key objectives of urban government. In place of a Marxist critique of the destruction of the experiential and affective life of the city, then, a more fruitful line of inquiry is to pursue a critique of the contellations of power, authority and truth through which the life of the city was defined, known and controlled.

## The Life of the City

In much writing on the avant-garde attempt to eliminate the boundary between art and life, there is an implicit assumption that 'life' is simply a passive material to be styled through art. Michel Foucault's histories of the 'biopolitical' constitution of modern experience, however, invite us to take fuller account of the role of biological life, as both object and objective, in the formation of modern spatial arts practices. In fact, the nature and dynamics of life as a process of growth, creativity and experiential intensity was itself a crucial problem of nineteenth-century government. This means that understanding the avant-gardist attempt to creatively transform life through art demands a new understanding of the ways in which problematizing the nature of art in modernity also meant intervening in biopolitical discourses and experiences. Montmartre bohemians, we will see, attempted to transform art through life just as much as they transformed life through art.

Over the course of the nineteenth century, the city emerged as a prominent biopolitical space, a vast laboratory for experimenting with new ecologies of human life. Through the research spawned by Foucault's enormously fertile accounts of the biopolitical constitution of modern experience, we know some of the ways in which biological life became both the object of power and also a key value and objective of power during the late eighteenth and nineteenth centuries.[48] In *The Order of Things,* Foucault described life, along with labour and language, as one of the three 'quasi-transcendentals' (that is, the condition of possibility of experience) of modern human science.[49] In his later work he analysed 'the attempt, starting from the eighteenth century, to rationalize the problems posed to governmental practice by phenomena characteristic of a group of living beings forming a population: health, hygiene, birth rate, life expectancy, race'.[50] Foucault even went so far as to refer to the emergence of biopolitical discourses as a 'threshold of modernity', where 'the life of the species is wagered on its own political strategies ... modern man is an animal whose politics places his existence as a living being in question'.[51]

Crucially, however, life was not simply the object of government, but also its objective. Modern government came to focus on ways of maximizing the life of the population. In modernity, life and those values associated with life – vitality, growth, energy, creativity – became 'the supreme standard and the highest good to which everything else is referred'.[52] Biopolitics 'takes life as both its object and its objective ... its basic function is to improve life, to prolong its duration,

---

48 On the spatial history of biopolitics, see, for example, Alison Bashford, 'Global Biopolitics and the History of World Health', *History of the Human Sciences* 19(1) (2006), Matthew Gandy, 'Zones of Indistinction: Bio-Political Contestations in the Urban Arena', *Cultural Geographies* 13 (2006), Matthew Hannah, *Governmentality and the Mastery of Territory in Nineteenth-Century America* (Cambridge, 2000), Margo Huxley, 'Spatial Rationalities: Order, Environment, Evolution and Government', *Social & Cultural Geography* 7(5) (2006), Stephen Legg, *Spaces of Colonialism: Delhi's Urban Governmentalities* (Oxford, 2007), David Nally, '"That Coming Storm": The Irish Poor Law, Colonial Biopolitics, and the Great Famine', *Annals of the Association of American Geographers* 98(3) (2008), Thomas Osborne, 'Security and Vitality: Drains, Liberalism and Power in the Nineteenth Century', in Andrew Barry, Thomas Osborne, and Nikolas Rose (eds), *Foucault and Political Reason: Liberalism, Neo-Liberalism, and Rationalities of Government* (Chicago, 1996), Thomas Osborne and Nikolas Rose, 'Governing Cities: Notes on the Spatialisation of Virtue', *Environment and Planning D: Society and Space* 17(6) (1999), Thomas Osborne and Nikolas Rose, 'Spatial Phenomenotechnics: Making Space with Charles Booth and Patrick Geddes', *Environment and Planning D: Society and Space* 22(2) (2004), Paul Rabinow, *French Modern: Norms and Forms of the Social Environment* (1989), R. Rose-Redwood, 'Governmentality, Geography, and the Geo-Coded World', *Progress in Human Geography* 30(4) (2006).

49 Michel Foucault, *The Order of Things: An Archaeology of the Human Sciences* (London, 1970).

50 Foucault, *The Birth of Biopolitics: Lectures at the Collège de France, 1978–1979*, p. 317.

51 Michel Foucault, *The Will to Knowledge: The History of Sexuality, Volume One*, 2nd ed. (London, 1998), pp. 142–3.

52 Johanna Oksala, 'Violence and the Biopolitics of Modernity', *Foucault Studies* 10 (2010), p. 25.

to improve its chances, to avoid accidents, and to compensate for its failings'.[53] In biopolitical reason, life becomes an end in itself, the ultimate value to be protected and fought for. Thus biopolitical logics did not just support oppressive forms of control and domination, but also revolutionary movements and the founding of the welfare state.[54]

As Foucault observes, and subsequent analysts have drawn out in more detail, a privileged site of experimentation with the life of the population was the city.[55] Over the course of the nineteenth century, a way of understanding and governing the city became predominant where the city became theorized as an organic milieu that was to be controlled ecologically rather than mechanically or deterministically. Participating in certain currents of Romanticism, which aspired to achieve a harmony between mankind and the natural world, this discourse saw the city essentially as 'an immanent naturalistic domain; one which, left to itself and with the right conditions, could compose itself as a benign social order'.[56] If an increasing priority for government was to secure the health of the population, and society was to be evaluated according to its health and vitality, then securing an appropriate environment that could foster life was a crucial way of ensuring the vitality of the population as a whole.

Following the dramatic renovations of Parisian infrastructure during the Second Empire, the naturalization of the city continued and evolved during the Third Republic.[57] The impacts of the concern for the health of the urban environment were very widespread.[58] For example, it was the renewed emphasis on the natural environment that led to vigorous efforts to combat the problem

---

53   Michel Foucault, *Society Must Be Defended*, trans. D. Macy (London, 2004), p. 254.

54   Matthew Hannah, 'Biopower, Life and Left Politics', *Antipode* 43(4) (2011).

55   Michel Foucault, *Security, Territory, Population: Lectures at the Collège de France, 1977–1978*, trans. Graham Burchell (Basingstoke & New York, 2007). See also, for example, Huxley, 'Spatial Rationalities: Order, Environment, Evolution and Government', Legg, *Spaces of Colonialism: Delhi's Urban Governmentalities*, Osborne and Rose, 'Governing Cities: Notes on the Spatialisation of Virtue', Osborne and Rose, 'Spatial Phenomenotechnics: Making Space with Charles Booth and Patrick Geddes', Rabinow, *French Modern: Norms and Forms of the Social Environment*.

56   Osborne and Rose, 'Governing Cities: Notes on the Spatialisation of Virtue', pp. 741–2.

57   Francoise Choay, *The Modern City: Planning in the 19th Century*, trans. George R. Collins and Marguerite Hugo (London, 1969), David Harvey, *Paris: Capital of Modernity* (New York, 2003), François Loyer, *Paris Nineteenth Century: Architecture and Urbanism* (New York, 1988), David Pinkney, *Napoleon III and the Rebuilding of Paris* (Princeton, NJ, 1958), David van Zanten, *Building Paris: Architectural Institutions and the Transformation of the French Capital 1830–1870* (Cambridge, 1994).

58   Andrew Aisenberg, *Contagion: Disease, Government, and the 'Social Question' in Nineteenth-Century France* (Stanford, CA, 1999), Lion Murard and Patrick Zylberman, *L'Hygiène dans la République: La Santé Publique en France, ou, L'Utopie Contrariée, 1870–1918* (Paris, 1996).

of overcrowded proletarian housing, since there was a deep concern about the pathological environmental effects of urban slums.[59] Similarly, the renovation of the Paris water system was precipitated by concerns about the malign effects of bacteria and germs on urban health.[60] Phenomena that were labelled 'social diseases' such as prostitution and alcoholism were subjected to scrutiny that at times came close to hysteria.[61] Consequently, the city's cafés and cabarets, long a haven of proletarian sociability, were demonized as milieus of alcoholism, vice and degeneracy.[62] Women bore the brunt of this process of naturalization, and feminist arguments for birth control, for example, were dismissed out of hand as a risk to population growth, as the nation's declining birth rate became a matter of feverish hand wringing.[63] Despite all of these measures, however, a deep pessimism pervaded fin-de-siècle France, emerging from a sense that the nation had exhausted its energies and allowed itself to degenerate into an enervated population of sickly weaklings who would never regain the strength and vigour to avenge their humiliation by Prussia in 1870.[64]

As Paul Rabinow has shown, these issues were fundamental in the transformation of discourses surrounding the nature of the city during the nineteenth and early twentieth centuries.[65] As the modern notion of 'society' as a whole way of life, rather than 'high' society, developed from its initial formulation within the life sciences, leading to the emergence of the 'social question' in the last two decades of the nineteenth century, the city came to be understood in neo-Lamarckian terms as an ecological milieu, a natural environment for the evolution of a human population. When people were no longer regarded as isolated individuals but part of a social whole, governmental intervention came to focus on the most important milieu of that whole – the city. The result of this was the city's emergence as a privileged object of technical intervention, as it

---

59　Ann-Louise Shapiro, *Housing the Poor of Paris* (Madison, 1985).

60　David Barnes, *The Great Stink of Paris and the Nineteenth-Century Struggle against Filth and Germs* (Baltimore, 2006), Matthew Gandy, 'The Paris Sewers and the Rationalization of Urban Space', *Transactions of the Institute of British Geographers* 24(1) (1999).

61　Alain Corbin, *Women for Hire: Prostitution and Sexuality in France after 1850* (Cambridge, MA, 1990), Patricia Prestwich, *Drink and the Politics of Social Reform: Antialcoholism in France since 1870* (Palo Alto, 1988).

62　W. Scott Haine, *The World of the Paris Cafe: Sociability among the French Working Class, 1789–1914* (Baltimore, 1996).

63　Fabrice Cahen, 'Medicine, Statistics, and the Encounter of Abortion and "Depopulation" in France (1870–1920)', *History of the Family* 14 (2008), Karen Offen, 'Depopulation, Nationalism, and Feminism in Fin-De-Siècle France', *The American Historical Review* 89(3) (1984).

64　Charles Bernheimer, *Decadent Subjects: The Idea of Decadence in Art, Literature, Philosophy, and Culture of the Fin-De-Siècle in Europe* (Baltimore, 2001), Daniel Pick, *Faces of Degeneration: A European Disorder, C. 1848–C. 1918* (Cambridge, 1989).

65　Rabinow, *French Modern: Norms and Forms of the Social Environment*.

became the object of the expertise of engineers and urban planners, rather than of the aesthetic sensibilities of architects in the Beaux-Arts tradition. This eventually led to modern urbanism – a vision of the planned city as regulator of modern society. The 'life of the city' – the circulation of bodies, diseases, crime, vice, degeneration, destitution and delinquency – became a crucial focus of nineteenth century government.[66] It also became a vehicle through which to depoliticize government, making government a technical exercise rather than a political process. Such moves even made it possible for commentators to depoliticize one of the great revolutionary events of the nineteenth century, the Paris Commune. The Commune, wrote Jules Claretie, was 'more pathological than political. The excessive cerebral agitation of the preceding months burst out in an immense fit'.[67]

However, the life of the city also became an important point of contestation in late nineteenth century urban culture. It is a specific set of cultural interventions in the life of the city that is the principal object of investigation of this study. Here it is necessary to move beyond the focus of much Foucauldian literature on the technical administration of life. A hugely important literature on urban bipolitics has emphasized the growing role of the technical expertise of engineers, urban planners, doctors, architects, statisticians and cartographers, in the government of the city. Yet the life of the city was not only a technical object of experimentation, but also a *cultural* and *performative* space of experimentation. Experiments with the life of the city were also conducted by artists, writers and musicians. Moreover, since the city was a privileged site of intersection of art, the everyday, and biological life, it was the city that became a key focus of avant-garde artistic and political experiments. This book advances historical and theoretical debates concerning the politics of life in the modern city, therefore, by investigating the role of arts and popular culture in testing, contesting, and reworking the forms of biopolitical knowledge, authority and expertise through which the life of the city was governed.

## Life and the Experience of Modernity

> There is a mode of vital experience – experience of space and time, of the self and others, of life's possibilities and perils – that is shared by men and women all over the world today. I will call this body of experience 'modernity'. To be modern is to find ourselves in an environment that promises us adventure, power, joy, growth, transformation of ourselves and the world – and, at the same time,

---

66   Osborne and Rose, 'Governing Cities: Notes on the Spatialisation of Virtue', p. 741.

67   'L'état de Paris était encore plus pathologique que politique. La surexcitation cérébrale des derniers mois éclatait en un immense accès'. Jules Claretie, *Histoire De La Révolution De 1870–71* (Paris, 1972). Cited in Janet Beizer, *Ventriloquized Bodie: Narratives of Hysteria in Nineteenth-Century France* (Ithaca, 1994).

that threatens to destroy everything we have, everything we know, everything we are. Modern environments and experiences cut across all boundaries of geography and ethnicity, of class and nationality, of religion and ideology: in this sense, modernity can be said to unite all mankind. But it is a paradoxical unity, a unity of disunity: it pours us all into a maelstrom of perpetual disintegration and renewal, of struggle and contradiction, of ambiguity and anguish. To be modern is to be part of a universe in which, as Marx said, 'All that is solid melts into air'.[68]

Much has been written about the 'experience of modernity', and its roots in the changing flows of capital, since Marshall Berman's celebrated study.[69] In particular, the idea that there is a single, universally applicable experience of modernity has been strongly criticized for neglecting differences in class, gender, ethnicity, and other social divisions.[70] Here, however, in order to conceptualize the role of the arts in creatively intervening in the life of the city, I would like to dwell briefly on the role of biopolitical discourses in the constitution of modern experience.

Literature on biopolitical modernity has not always emphasized enough the extent to which the elevation of life as a supreme value motivated a parallel celebration of the energies of lived, embodied experience. Whilst Ben Anderson has persuasively argued that the exercise of biopower is deeply tied up with the politics of affect in contemporary capitalism, there is no reason to suppose that this is only true of contemporary biopower.[71] In fact, a characteristic attribute of modernity is the way in which intensified subjective experience emerged as the most dynamic and creative expression of life and vitality.[72] As Claire Blencowe writes, 'the history of biopolitics is *also* the history of the positivity – the experiential, embodied, perceptive and affective forces – of modernity and the formations

---

68    Marshall Berman, *All That Is Solid Melts into Air: The Experience of Modernity* (London, 1983), p. 1.

69    Some of the most influential include Zygmunt Bauman, *Liquid Modernity* (Cambridge, 2000), A. Giddens, *Modernity and Self-Identity: Self and Society in the Late Modern Age* (1991), David Harvey, *The Condition of Postmodernity: An Enquiry into the Origins of Cultural Change* (Oxford, 1989).

70    For example, Gurminder Bhambra, *Rethinking Modernity: Postcolonialism and the Sociological Imagination* (Basingstoke, 2007), Rosalyn Deutsche, 'Boys Town', *Environment and Planning D: Society & Space* 9 (1991), S. N. Eisenstadt, 'Multiple Modernities', *Daedalus* 129(1) (2000), Janet Wolff, 'The Invisible Flâneuse: Women and the Literature of Modernity', *Theory, Culture & Society* 2(3) (1985).

71    Ben Anderson, 'Affect and Biopower: Towards a Politics of Life', *Transactions of the Institute of British Geographers* 37(1) (2012).

72    This is one of the key points to be taken from Walter Benjamin's work on modern Paris. See Walter Benjamin, *The Arcades Project*, ed. R. Tiedmann, trans. K. McLaughlin and H. Eiland (Cambridge, MA, 1999), Walter Benjamin, 'On Some Motifs in Baudelaire', in Hannah Arendt (ed.), *Illuminations* (London, 1999), Susan Buck-Morss, *The Dialectics of Seeing: Walter Benjamin and the Arcades Project* (Cambridge, MA, 1989).

of value and evaluation that these entail. It is the history of a proliferation of experience, of contingency and creativity; a politicization of embodiment and life; and a production of life – vitality, creativity, health, reproduction, security and evolution – as *values*, inspiring and motivating ethical, epistemological and political work'.[73] The life of the city was not simply a technical object of intervention, but also something that expressed itself in directly experiential terms as a force of intensity, growth, and creativity.

By highlighting the role of arts and popular culture in the testing and contesting of biopolitical urban discourses, I will draw on an important body of work exploring the role of embodied experience in the operation of biopolitical techniques of power, and in particular on performative interventions on biopolitical rationalities.[74] However, my intention is not to suggest that there is any unified biopolitical experience of modernity. Rather, biopolitical discourses, and the valorization of life as a social and cultural value that they motivated, created a framework of possibilities through which it was possible to stylize experience through new arts of life.

Although Foucault's own work has sometimes been criticized for its neglect of experience and affect, some comments in his later work indicate that we might productively reread his oeuvre as a history of experience.[75] Foucault himself described his work as

---

73   Claire Blencowe, *Biopolitical Experience: Foucault, Power and Positive Critique* (Basingstoke, 2012), p. 3.

74   On performance, art and biopolitics, see for example, Christian Abrahamsson and Sebastian Abrahamsson, 'In Conversation With the Body Conveniently Known as Stelarc', *Cultural Geographies* 14 (2007), Beatriz da Costa and Kavita Philip (eds), *Tactical Biopolitics: Art, Activism, and Technoscience* (Cambridge, MA, 2008), Deborah Dixon, 'Creating the Semi-living: On Politics, Aesthetics and the More-than-human', *Transactions of the Institute of British Geographers*, 34 (2009), Derek McCormack 'Diagramming Practice and Performance', *Environment and Planning D: Society and Space*, 23 (2005), J.-D. Dewsbury, 'Affective Habit Ecologies: Material Dispositions and Immanent Inhabitations', in *Performance Research: A Journal of the Performing Arts*, 17 (2012), Peter Kraftl, 'Liveability and Urban Architectures: Mol(ecul)ar Biopower and the "Becoming Lively" of Sustainable Communities', *Environment and Planning D: Society and Space*, 32 (2014), Nigel Thrift, 'Lifeworld Inc.: And What to do About it', *Environment and Planning D: Society and Space*, 29 (2011). Alexander Vasudevan, 'Symptomatic Acts, Experimental Embodiments: Theatres of Scientific Protest in Interwar Germany', *Environment and Planning A* 39(8) (2007), Katherine Yusoff, 'Biopolitical Economies and the Political Aesthetics of Climate Change', *Theory, Culture & Society*, 27 (2010).

75   On the critique of Foucault's lack of attention to experience, see for example Lois McNay, 'The Foucauldian Body and the Exclusion of Experience', *Hypatia* 6(3) (1991), Nigel Thrift, 'Overcome by Space: Reworking Foucault', in Jeremy Crampton and Stuart Elden (eds), *Space, Knowledge and Power: Foucault and Geography* (Aldershot, 2007). For more positive accounts of the role of experience in Foucault's thought, see Blencowe, *Biopolitical Experience: Foucault, Power and Positive Critique*, Peter Hallward, 'The Limits of Individuation, or How to Distinguish Deleuze and Foucault', *Angelaki* 5(2)

an analysis of what could be called focal points [foyers] of experience in which forms of a possible knowledge (*savoir*), normative frameworks of behaviour for individuals, and potential modes of existence for possible subjects are linked together ... these three things, or rather their joint articulation, can be called, I think, a 'focal point of experience'.[76]

Conceptualizing experience in this way, as a joint articulation of knowledge, norms and ethics, makes it impossible to talk meaningfully about any single 'experience of modernity'. Rather, experience is constituted through ethical relationships towards the present, as subjects stablish new styles of living within the contemporary limits of knowledge and structures of normalization. Modernity is an ethos or way of styling experience within the present, rather than being reducible to a specific experiential content.[77] It is 'a mode of relating to contemporary reality; a voluntary choice made by certain people; in the end, a way of thinking and feeling'.[78]

For this reason, the 'art of living' can be viewed as a creative project involving testing the limits of experience. Reinventing the subject, Foucault writes, involves 'trying to reach a certain point in life that is as close as possible to the "unliveable", to that which can't be lived through'. Here, 'experience has the function of wrenching the subject from itself, of seeing to it that the subject is no longer itself, or that it is brought to its annihilation or its dissolution'.[79] Whereas lived experience, as a supposedly 'true' experience, establishes the unity of a subject's life, what Foucault calls a 'limit-experience' has precisely the opposite effect: it transfigures the limits of the subject, making perceptible new possibilities of conduct, and thereby interrupting the established boundaries of life. This kind of experience, far from being an expression of the 'the bubbling source of life itself, life in an as yet uncaptured state', is an experience that reaches towards the *outside* of life.[80]

The urban cultural experiments that I analyse in this book, then, can be interpreted as attempts to create 'limit experiences' that tested, not just the

---

(2000), Béatrice Han, *Foucault's Critical Project: Between the Transcendental and the Historical*, trans. E. Pyle (Stanford, CA, 2002), Martin Jay, *Songs of Experience: Modern American and European Variations on a Universal Theme* (Berkeley, CA, 2005), Leonard Lawlor, *The Implications of Immanence: Towards a New Concept of Life* (New York, 2006), Clare O'Farrell, *Foucault: Historian or Philosopher?* (Basingstoke, 1980).

76   Michel Foucault, *The Government of Self and Others* (Basingstoke, 2010), p. 3.
77   Michel Foucault, 'What Is Enlightenment?', in Paul Rabinow (ed.), *Ethics, Subjectivity and Truth: The Essential Works of Foucault 1954 -1984, Volume One* (London, 2000). On modernity as ethos, see Thomas Osborne, *Aspects of Enlightenment: Social Theory and the Ethics of Truth* (London, 1998).
78   Foucault, 'What Is Enlightenment?', p. 309.
79   Michel Foucault, 'Interview with Michel Foucault', in James Faubion (ed.), *Power. Essential Works of Foucault, 1954–1984, Volume Three* (London, 2002), p. 241.
80   Michel Foucault, *The Archaeology of Knowledge*, trans. A. M. Sheridan Smith, 2nd ed. (London & New York, 2002), p. 85.

experiential limits of the subject, but the experiential life of the city itself. Art and popular urban culture formed powerful practices through which the life of the city could be experienced, made visible, and reimagined. Through art, dominant truths concerning the vitality of social life could be challenged and new truths about the life of the city could be voiced.

## Authority and Experience

In his discussion of modernity as an ethos rather than an epoch or a type of experience, Foucault emphasizes the fact that an art of living demands styling a new relationship to authority. Foucault draws on Kant's essay 'What is Enlightenment?', which defines Enlightenment in terms of an emancipation from 'alien guidance'. Immaturity, Kant writes, involves lazy and unthinking obedience to external authority – 'a book to have understanding in place of me, a spiritual adviser to have a conscience for me, a doctor to judge my diet for me'.[81] Thus Enlightenment is defined by Kant as 'the moment when humanity is going to put its own reason to use, without subjecting itself to any authority'.[82] Or to be more accurate, we might say that Enlightenment is the moment where humanity does not subject itself to external authority, but is only guided by forms of *immanent* authority that are internal to the faculties of reason, imagination and understanding. Foucault's appeal to an art of living, similarly, gestures towards forms of subjectivity that do not refer to externally imposed standards of truth or morality, but are lived according to immanent, aesthetic judgments of beauty, style and freedom.[83]

As Foucault made clear in his lectures on Greek parrhesia or 'fearless truth-telling', however, contesting dominant truth claims requires generating new forms of authority through which to speak truth to power. As Nancy Luxon explains, parrhesia 'encompasses a broad set of personalized ethical practices that construct relationships to oneself, to authority, and to truth'.[84] Parrhesia is a speech act wherein people who do not speak from institutional positions of authority attempt to speak truth to power, and derive the authority to speak from their demonstration of courage, commitment and credibility. Over the course of this book I shall examine some ways in which the Montmartre avant-garde revived aspects of this

---

81    Immanuel Kant, 'What Is Enlightenment?', in L. Beck (ed.), *Selections* (New York, 1988).

82    Foucault, 'What Is Enlightenment?'.

83    For a more detailed statement of this argument, see Julian Brigstocke, 'Immanent Authority and the Performance of Community in Late Nineteenth Century Montmartre', *Journal of Political Power* 6(1) (2013).

84    Nancy Luxon, 'Truthfulness, Risk, and Trust in the Late Lectures of Michel Foucault', *Inquiry* 57(5) (2004), Nancy Luxon, *Crisis of Authority: Politics, Trust, and Truth-Telling in Freud and Foucault* (Cambridge, 2013).

ethos of courageous truth-telling in order to generate 'immanent' forms of authority to critique dominant structures of biopolitical knowledge and experience.

My argument will be that the creative community of artists and anarchists that came together in Montmartre provides an illuminating example of an influential historical attempt to stylize a new aesthetics of authority. Through cultural performance, including theatrical, everyday and textual performances – as well as the violent performances of anarchist 'propaganda by the deed' – the community attempted to find ways of making authoritative claims towards social, political and economic change. They did not only attempt to live out a utopian alternative lifestyle, I will argue, but to make this lifestyle an authoritative example for social and political transformation.

Consequently, my argument diverges from scholarly accounts that describe Montmartre's urban culture as a quintessentially anti-authoritarian one. John Munholland, for example, describes the area's 'long tradition of independence and defiance of all forms of authority'.[85] Similarly, Raymond Jonas describes it as a community 'suspicious of authority of all kinds'.[86] Certainly it is true that the new forms of urban culture that were developed there did pose a powerful challenge to traditional institutions of authority – including state, church, family, and the main institutions and monuments of the city itself. My argument, however, is that it would be more accurate to describe the intentions of the Montmartre community as being not just to destroy established forms of authority but to experiment with using embodied, affective experience as the grounds from which to engineer new forms of bottom-up authority. The aim of the Montmartre artistic community was to create a cultural explosion of such luminescence that the rest of the world would be forced to take notice. The aims of this group were not merely destructive, but also *con*structive, in that they were committed to creating a wholly new form of aesthetic community that would provide the foundation for new values, new styles of living, and new political formations.

## Spatialities of Humour

Perhaps the most distinctive technique through which Montmartre bohemians attempted to stylize new forms of authority, truth and experience was a carnivalesque variety of humour. Montmartre quickly became known for a style of humour termed 'fumisme', roughly translatable as 'blowing smoke'. Fumisme, as Michael Wilson notes, can be seen as 'a more subtle outgrowth of *la blague*, the

---

85   John Kim Munholland, 'Republican Order and Republican Tolerance in Fin-De-Siècle France: Montmartre as Delinquent Community', in Gabriel Weisberg (ed.), *Montmartre and the Making of Mass Culture* (New Brunswick, NJ, 2001), p. 22.

86   Raymond Jonas, 'Sacred Tourism and Secular Pilgrimage: Montmartre and the Basilica of Sacré Coeur', in Gabriel Weisberg (ed.), *Montmartre and the Making of Mass Culture* (New Brunswick, NJ, 2001), p. 110.

joking hyperbole and buffoonery of the artist's studio ... Like la blague, though, the bohemians' fumisme gains its power by forgoing direct attacks and cloaking its scorn in a genial manner'.[87]

In recent years, humour has become the object of lively social scientific debate. Rejecting the assumption that there is anything inherently positive, progressive, or anti-authoritarian in humour, research on humour has brought out the ways in which jokes can act as a powerful affective vehicles for creating and consolidating boundaries: securing, in other words, the integrity of the inside and outside of social groupings. Scholars have discovered numerous ways in which humour has the power to include and exclude, since in order to work it relies upon nuanced sensitivities to shared histories, traditions or codes.[88] A shared sense of humour is both a powerful coping mechanism and a highly effective social bond, as well as means of naturalizing learned differences, and hence, social hierarchies. As well as consolidating group boundaries and identities, however, humour can also disrupt them, and scholars have uncovered a variety of ways in which humour can challenge established representations and operate as a form of creative resistance or subversion.[89] Discursive analysis of humour has proved highly effective in uncovering the ways in which humour operates in the production, reproduction and contestation of hegemonic identities and representations.

What academic scholarship on humour has paid less attention to, however, are the non-representational aspects of humour: the means by which humour moves us in ways that are not reducible to their semantic or discursive content. Humour, as many commentators have observed, is impossible to capture through representation. Georges Bataille went so far as to argue that 'That which is laughable may simply be the "*unknowable*" ... *The unknown makes us laugh*'.[90] Empirical instances of humour 'continually exceed the theoretical analysis one is able to give of them –

---

87    Michael Wilson, 'Portrait of the Artist as a Louis XIII Chair', in Gabriel Weisberg (ed.), *Montmartre and the Making of Mass Culture* (New Brunswick, NJ, 2001), p. 182.

88    Michael Billig, *Laughter and Ridicule: Towards a Social Critique of Humour* (London, 2005), Mary Jane Kehily and Anoop Nayak, '"Lads and Laughter": Humour and the Production of Heterosexual Hierarchies', *Gender and Education* 9(1) (1997), Giselinde Kuipers, *Good Humor, Bad Taste: A Sociology of the Joke* (Berlin & New York, 2006), Sharon Lockyer and Michael Pickering (eds), *Beyond a Joke: The Limits of Humour* (Basingstoke, 2005), Chris Powell and George Paton (eds), *Humour in Society: Resistance and Control* (Basingstoke, 1988), Teela Sanders, 'Controllable Laughter: Managing Sex Work through Humour', *Sociology* 38(2) (2004), Simon Weaver, *The Rhetoric of Racist Humour: US, UK and Global Race Joking* (Farnham, 2011), Paul Willis, *Learning to Labour: How Working Class Kids Get Working Class Jobs* (Aldershot, 1977).

89    See, for example, Marjolein t'Hart and Dennis Bos (eds), *Humour and Social Protest* (Cambridge, 2007), Simon Weaver, 'The "Other" Laughs Back: Humour and Resistance in Anti-Racist Comedy', *Sociology* 44(1) (2010).

90    Georges Bataille and Annette Michelson, 'Un-Knowing: Laughter and Tears', *October* 36 (1986), p. 90. Emphasis in the original. See Lisa Trahair, 'The Comedy of Philosophy: Bataille, Hegel and Derrida', *Angelaki* 6(3) (2001).

they say more in saying less'.[91] There is a powerful corporeal, embodied, affective element to humour whose political dynamics need to be understood more clearly. Maria Hynes and Scott Sharpe, for example, emphasize the need to approach humour in ways that do not simply judge it through a representational logic that ties it to a pre-existing moral framework, but that remain attentive to the new openings that it can create.

> [W]hile judgments of meaning and value may wrest from humour something useful, they risk denying the rich actuality and the potential of what they encounter, since their judgment point is an already given, ideal image of the world ... If humour can teach us something, it is not because it is a pliable medium of moral didacticism. Rather, it may serve as an entry point into understanding our desires, their social formation and the ways they translate as moral justifications. Such a genealogical enquiry is already affirmative, because it opens the way to a spirited revaluation of values.[92]

Developing an affirmative stance towards humour demands remaining attentive to its corporeal dynamics, its ability to affect bodies in ways that are not reducible to its discursive content. Indeed, I shall demonstrate in this book that understanding the affective qualities of humour requires attending to its imbrication in the biopolitical dynamics of life and lived experience. Since the nineteenth century, we will see, humour has been closely tied to vitalist theories of embodied experience.

In addition, the book will excavate aspects of the *spatial* politics of humour. The spatial aspects of humour have received relatively little attention in the scholarly literature. There are several ways, however, in which humour plays a role in the production and contestation of space and place. For example, because humour is an effective means of stabilizing insides and outsides, humour – or its exclusion – can be an important device for creating and controlling 'affective atmospheres'.[93] In addition, humour, satire, and the sometimes violent debates they trigger, can play important roles in geopolitical disputes.[94] Humour is also used in

---

91    Simon Critchley, *On Humour* (London, 2001), p. 66.
92    Maria Hynes and Scott Sharpe, 'Yea-Saying Laughter', *Parallax* 16(3) (2010), pp. 51–2. See also Maria Hynes, Scott Sharpe and Bob Fagan, 'Laughing with the Yes Men: The Politics of Affirmation', *Continuum* 21(1) (2007), Scott Sharpe, Maria Hynes and Robert Fagan, 'Beat Me, Whip Me, Spank Me, Just Make It Right Again: Beyond the Didactic Masochism of Global Resistance', *Fibreculture* 6 (2005).
93    David Bissell, Maria Hynes and Scott Sharpe, 'Unveiling Seductions Beyond Societies of Control: Affect, Security, and Humour in Spaces of Aeromobility', *Environment and Planning D: Society & Space* 30 (2012), Lauren L. Martin, 'Bombs, Bodies, and Biopolitics: Securitizing the Subject at the Airport Security Checkpoint', *Social & Cultural Geography* 11(1) (2010).
94    Klaus Dodds and Philip Kirby, 'It's Not a Laughing Matter: Critical Geopolitics, Humour and Unlaughter', *Geopolitics* 18(1) (2013), Daniel Hammett, 'Zapiro and Zuma: A Symptom of an Emerging Constitutional Crisis in South Africa?', *Political Geography* 29(2)

attempts to subvert and re-politicize urban space.[95] Finally, humour can be used as a powerful technique for creating new 'ecologies of experience' that enable new forms of place and community to emerge.[96] As Hannah Macpherson argues, 'Jokes and humour often reveal certain assumptions about the received norms and values of certain places and people, and can sometimes be indicative of who is considered "in place" and who is "out of place"'.[97] Indeed, Montmartre's distinctive brand of humour, I will argue, was one of the most important ways through which a distinctive sense of place and marginality was created there. Humour emerged as a form of affective experience that could successfully express the contradictory aspects of modern experience in Montmartre. Moreover, in contrast to the usual emphasis on the anti-authoritarian aspects of humour, I will argue that part of the appeal of humour to the fin-de-siècle avant-garde was that it offered a possible ground for new forms of authority through which hegemonic ideas concerning the life of the city could be tested. Some anarchist radicals, for example, found in humour a way of lending authority to their claims to embody the most dynamic, vital and lively forms of urban politics (see Chapter 8).

To begin exploring this argument, we will start with an aesthetic figure that was to prove hugely influential in the emergence of an avant-garde bio-cultural politics in Montmartre. This was the image of the black cat: a figure that crystallized key aspects of the biopolitical experience of truth in nineteenth century French culture.

---

(2010), Daniel Hammett, 'Resistance, Power and Geopolitics in Zimbabwe', *Area* 43(2) (2011), Martin, 'Bombs, Bodies, and Biopolitics: Securitizing the Subject at the Airport Security Checkpoint', Darren Purcell, Melissa Scott Brown and Mahmut Gokmen, 'Achmed the Dead Terrorist and Humor in Popular Geopolitics', *GeoJournal* 75 (2010), Juha Ridanpää, 'Geopolitics of Humour: The Muhammed Cartoon Crisis and the Kaltio Comic Strip Episode in Finland', *Geopolitics* 14(4) (2009).

95   Kate Epstein and Kurt Iveson, 'Locking Down the City (Well, Not Quite): Apec 2007 and Urban Citizenship in Sydney', *Australian Geographer* 40(3) (2009), Anja Kanngieser, *Experimental Politics and the Making of Worlds* (Farnham, 2013),
    Paul Routledge, 'Sensuous Solidarities: Emotion, Politics and Performance in the Clandestine Insurgent Rebel Clown Army', *Antipode* 44(2) (2012).

96   Julian Brigstocke, 'Defiant Laughter: Humour and the Aesthetics of Place in Late 19th Century Montmartre', *Cultural Geographies* 19(2) (2012), Tim Cresswell, 'Laughter and the Tramp', *The Tramp in America* (London, 2013), Kevin McHugh and Ann Fletchall, 'Festival and the Laughter of Being', *Space and Culture* 15(4) (2012). On ecologies of experience, see Paul Simpson, 'Ecologies of Experience: Materiality, Sociality, and the Embodied Experience of (Street) Performing', *Environment and Planning A*, 45, 2013.

97   Hannah Macpherson, '"I Don't Know Why They Call It the Lake District. They Might as Well Call It the Rock District!" The Workings of Humour and Laughter in Research with Members of Visually Impaired Walking Groups', *Environment and Planning D: Society and Space* 26(6) (2008), p. 1082.

# PART I
# The Life of the City

# Chapter 2

# Anomalous and Inhuman: Black Cats and the Experience of Truth in Baudelaire, Manet and Poe

## Introduction

The explosion of Montmartre onto the cultural map of Paris started, as we have seen, at the Chat Noir 'cabaret artistique' on the old rue Victor Massé. Over the entrance was a sign of a black cat by a golden sun and an inscription reading, 'Passer-by, Halt! Be modern!' (Figure 2.1). Soon the figure of the black cat became an ubiquitous signifier for the distinctively Montmartrois take on modern urban living. Later it would become an important part of the iconography of twentieth century anarchism and revolutionary socialism, via the black cat of the anarcho-syndicalist movement and then the Black Panther black power movement (Figure 2.2). But what was the significance of the black cat, and why did this figure so capture the imagination of Montmartre artists and anarchists? In this chapter, we shall explore the rich field of iconographic meanings that the black cat possessed in 1880s Paris, in particular through a reading of the significance of the black cat in works by Edgar Allen Poe, Charles Baudelaire, and Édouard Manet. The black cat, I will suggest, emerged during the nineteenth century as a figure representing a distinctively modern experience of truth: a truth existing, not safely within the interior of the human subject, but at the outer limits of human life. By the 1880s, the black cat had become the quasi-mystical signifier of unrepresentable forms of life, experience and truth beyond the boundaries of the human subject.

## Edgar Allan Poe's 'The Black Cat'

On 27 January 1847, Charles Baudelaire was leafing through the new volume of *La Démocratie Pacifique*. He came across Isabelle Meunier's translation of Edgar Allan Poe's macabre short story 'The Black Cat', and was stunned by Poe's ability to express the strangeness of beauty and the fascination of evil, qualities he believed to capture perfectly elements of his own flourishing literary sensibilities. 'I have found an American author', he wrote to his mother, 'who has roused in me an incredible sympathy, and I have written two articles on his life and works.

**Figure 2.1** The Chat Noir. In Marcus du Seigneur, 'Vitruve et Gambrinus et le Chat Noir', *La Construction Moderne* (1885): 517–19

**Figure 2.2**   Evolution of the black cat as symbol of radical left-wing politics. Left: poster for the touring Chat Noir shadow theatre, 1890s. Centre: the 'wild cat' or 'sabot cat', symbol of the twentieth century anarchist movement. Designed by Ralph Chaplin, a prominent member of the Industrial Workers of the World, in the early 1900s. Right: Poster for the Black Panther Party, 1970s

They are written with ardour; but you will doubtless discover there some signs of a very extraordinary overexcitement. It is the consequence of the sad and mad life that I lead'.[1] What so astonished Baudelaire was 'the intimate resemblance, although not positively accented, between my own poems and those of this man'.[2]

Poe is perhaps most well known for his invention of the detective story, a distinctively modern form of urban narrative that captured an emerging experience of the big city as something alien and illegible, a space of crime and mystery, requiring a heroic effort to distinguish and decipher the forest of clues in the urban fabric. The detective story was a genre that appealed to the middle classes and evoked an inhospitable urban environment filled with danger, fear, and unease. '[B]ourgeois society always feels under attack; political crisis, social crisis, ideological terror are its permanent state of existence; therefore we always play

---

1   'J'ai trouvé un auteur américain qui a excité en moi un incroyable sympathie, et j'ai écrit deux articles sur sa vie et ses ouvrages. C'est écrit avec ardeur; mais tu y découvriras sans doute quelques signes d'une très extraordinaire surexcitation. C'est la conséquence de la vie douloureuse et folle que je mène'. Cited in P. Mansell Jones, 'Poe and Baudelaire: The "Affinity"', *The Modern Language Review* 40(4) (1945), p. 280.
2   Ibid.

the detective – and read detective fiction'.[3] Poe's detective Dupin possesses the rare capacity to uncover the truth in dark, shadowy crevasses that remain opaque to the ordinary urban dweller.

Poe's 'confessional' stories, however – of which 'The Black Cat' is perhaps the most important example – explore a different (though related) experience of truth in modernity. The story is one of his most disturbing, drawing on a preconscious world of elemental violence that threatens to rupture the orderly existence of the everyday world. It is told from a space at the outer limit of life – the narrator is in prison, shortly to suffer the hangman's noose. It evokes the bristling terror of uncontrollable forces, a feeling for nature's excess over human understanding. The world Poe evokes in this story is dark, obscure, and incomprehensible; it is a world in which reactive passions, fuelled by intoxication, circulate unrestrained, far exceeding the bonds of reason. The narrator, a warm-hearted lover of animals, owns (amongst other pets) a large cat called Pluto, who is black, intelligent, and strikingly beautiful.[4] However, his good character distorted by the 'disease' of alcohol, the narrator starts to ill-treat the animal. One night, on returning from a night on the town, 'the fury of a demon' possesses him and when Pluto scratches his hand, 'a more than fiendish malevolence' takes hold of him.[5] Drawing a knife from his pocket, the narrator attacks the cat, cleaving one of its eyes from its socket. The cat recovers, albeit now with a frightful appearance. But the narrator soon becomes compelled, by an 'unfathomable longing of the soul ... to do violence to its own nature' to consummate the injury he had inflicted earlier.[6] He slips a noose around the cat's neck and hangs it from the branch of a tree outside his house.

The night of this cruel act, the narrator's house burns to the ground. Inspecting the ruins, he finds the figure of a giant cat, with a rope around its neck, burned upon a piece of white plaster that had survived the conflagration. Although he finds a rational explanation for this appearance, the narrator is consumed with regret, unable to rid himself of its phantom. Going so far as to search for another pet, he comes across another black cat, also one-eyed, while spending some time 'in a den of more than infamy'. No-one knows where the cat came from, and he takes it home as a pet. It quickly domesticates itself, and becomes cloyingly affectionate. But soon the narrator starts 'to look upon it with unutterable loathing,

---

3    Carlo Salzani, 'The City as Crime Scene: Walter Benjamin and the Traces of the Detective', *New German Critique* 34(1) (2007), p. 169.

4    The name Pluto refers to the chthonic Greek god of the underworld, who shared sovereignty over the world with Zeus (ruler of the heavens) and Poseidon (master of the sea). Thus it immediately evokes a world of underground spirits, demons and uncanny presences. The denouement of the story takes place underground, in the narrator's cellar.

5    Edgar Allan Poe, 'The Black Cat', *The Raven and Other Poems and Tales* (Boston, New York & London, 2001), p. 28.

6    Ibid.

and to flee silently from its odious presence, as from the breath of a pestilence'.[7] As his 'absolute dread' for the beast increases, its affection only increases.

Soon, the narrator is moved by fury to kill the cat with an axe. But when his wife intervenes to stay the blow, in a fit of rage he launches the axe at his wife's own head, and kills her dead. Now, 'with the hideous murder accomplished, I set myself forthwith and with entire deliberation, to the task of concealing the body'.[8] He hits upon the idea of concealing the body in the cellar by building a false wall. When a party of policemen comes to search the house, however, just as they have finished searching the cellar for the third or fourth time and after the narrator knocks on the wall in barely suppressed triumph, they hear a terrible cry – 'at first muffled and broken, like the sobbing of a child, and then quickly swelling into one long, loud, and continuous scream, utterly anomalous and inhuman – a howl – a wailing shriek, half of horror and half of triumph, such as might have arisen only out of hell, conjointly from the throats of the damned in their agony and of the demons that exult in the damnation'. The wall is demolished and the corpse is found, putrifying and covered with gore; and 'Upon its head, with red extended mouth and solitary eye of fire, sat the hideous beast whose craft had seduced me into murder, and whose informing voice had consigned me to the hangman. I had walled the monster up within the tomb'.[9]

The crime is motiveless, a transgression of the social bond that cannot be rationally explained. Despite the narrator's careful and cold-hearted rationalizing of the best means to conceal a body, the murder itself has no rational end. It is a response to an overpowering experience of dread. The final image of the story, however – the 'red extended mouth and solitary eye of fire' – testifies to a double encounter with truth, and it is possible to read the story in terms of its evocation of a distinctive experience of truth in modernity. The story is *confessional* in a double sense. Firstly, the narrator confesses his crime to the reader. Secondly, however, he is responsible for his own discovery and seems to have an unconscious desire for his crime to be revealed. It is his triumphalist knocking on the false wall that prompts the ghastly howls of the immured cat. The narrator cannot contain his guilty secret, but is compelled by a desire to confess the truth.

By the end of the story, we know the truth. But we also know what path has to be travelled in order to experience it. Truth emerges, not through the power of human reason in the clear light of day, but hiding behind bricked walls in a space saturated with blood, calling for revenge, howling with anguish. To encounter the truth, it is necessary to descend into the underworld, a land of demons, spirits, and the living dead. The experience of truth is an experience of the limit of life, of reason, of humanity itself. Thus for the narrator, death – both his wife's and his own – is the necessary price of his determination to tell the truth. Truth emerges, not through the revelatory power of light, but through a sound whose source lies

---

7  Ibid., 31.
8  Ibid., 34.
9  Ibid., 38.

far behind the false wall of appearance and which pierces the metaphorical wall of the human body. Truth comes from a deeply felt, affective experience at the limits of organic life.

In this story, the black cat stands in for the malevolent side of unknowable supernatural forces. The cat haunts and unhinges the narrator, who blames it for the murderous act that leads him to the gallows.[10] Here the black cat is an aesthetic figure of the limit between life and death, and reveals the duplicitous nature of modern subjectivity. It is always double, both materially (appearing in two physical bodies) and emotionally, at once overbearingly intimate and disdainfully aloof, seductive and repellent, affectionate and spiteful, innocent and vengeful. It is a material incarnation of immaterial, occult forces. Although the reality of these forces is left in question – the narrator challenges the reader to come up with a rational explanation of the events he relates – the cat embodies a fear of the beyond and the unknown, an alarm at the frailties and weaknesses of human reason, a foreboding of the violence of the human psyche. It evokes forces that are 'anomalous and inhuman', inexplicable and unnatural. It testifies to the pathos of the human condition – that in order to discover the truth of life, it will be necessary to transgress its limits and risk death. The truth of life comes from an impenetrable, invisible, unrepresentable realm beyond its borders.

## Baudelaire's Cats

'Never has the truth of speech, a higher form of truth, shown itself so clearly … Truth, however, before it is rest, is a long violence'. So writes Yves Bonnefoy in a beautiful essay on Charles Baudelaire's *Les Fleurs du Mal* (Flowers of Evil).[11] Cats, I wish to suggest, played an important role in Baudelaire's exploration of the long violence of truth. Baudelaire, as we have seen, was fascinated by Poe's story, which he translated into French. The archetypal poet of the modern metropolis and a towering influence on fin-de-siècle Montmartre urban culture, Baudelaire was famous for his love of cats. This was during a time in which cats were considered highly undesirable household pets. Unlike dogs, who could be happily integrated into the bourgeois home, cats were associated with an intransigent and contemptible refusal to submit themselves fully to middle class domesticity.[12] Whereas the dog

---

10 As several commentators on the story have observed, however, the narrator's testimony is extremely unreliable. See, for example, James Gargano, 'The Question of Poe's Narrators', *College English* 25 (1963). Moreover, the narrator does not attest to the reality of these occult forces, but leaves it to the reader to make her own judgments about the true cause of the events.

11 Yves Bonnefoy, 'Baudelaire's *Les Fleurs Du Mal*', in John Naughton (ed.), *The Act and the Place of Poetry: Selected Essays* (Chicago, 1989), pp. 44–5.

12 Kathleen Kete, *The Beast in the Boudoir: Petkeeping in Nineteenth-Century Paris* (Berkeley, CA, 1994).

was associated with the home, the cat was associated with the street. Baudelaire described his cat as 'the sole source of amusement in [his] lodgings', adding that 'the sight of dogs sickens me'.[13] He devoted no fewer than three poems to cats in *Les Fleurs du Mal*, as well as a prose poem in *Le Spleen de Paris*, and cats make appearances in many other of his poems.

The poet Théophile Gautier, to whom Baudelaire dedicated his cycle of poems, offered an explanation for Baudelaire's fascination with cats:

> [I]n these sweet animals there is a nocturnal side, mysterious and cabalistic, which was very attractive to the poet. The cat, with its phosphoric eyes, which are like lanterns and stars to him, fearlessly haunts the darkness, where he meets wandering phantoms, sorcerers, alchemists, necromancers, resurrectionists, lovers, pickpockets, assassins, grey patrols, and all the obscene spectres of the night. He appears to know the latest sabbatical chronicle, and he will willingly rub himself against the lame leg of Mephistopheles. His nocturnal serenades, his loves on the tiles, accompanied by cries like those of a child being murdered, give him a certain satanic air which justifies up to a certain point the repugnance of diurnal and practical minds, for whom the mysteries of Erebus do not have the slightest attraction.[14] But a doctor Faustus, in his cell littered with books and instruments of alchemy, would always love to have a cat for a companion.[15]

For Gautier, then, cats were a bestial, other-worldly incarnation of that famous nineteenth-century urban persona the 'flâneur', the heroic urban stroller who was so central to Baudelaire's vision for modern art and life.[16]

Baudelaire's cat poems have proved fertile ground for literary theory, most famously Roman Jakobson and Claude Levi-Strauss's structuralist reading of 'Le Chat'.[17] What I want to explore here, however, is a more specific issue

---

13    Rosemary Lloyd (ed.), *Selected Letters of Charles Baudelaire: The Conquest of Solitude* (Chicago, 1986), p. 274.

14    Erebus, meaning 'deep darkness' or 'shadow' in Ancient Greek, was the region of the Underworld where the dead passed immediately after dying.

15    Théophile Gautier, *Charles Baudelaire: His Life*, trans. Guy Thorne (New York, 1915), p. 42.

16    On the flâneur, see, in addition to Walter Benjamin's seminal works (for example Walter Benjamin, *Charles Baudelaire: A Lyric Poet in the Era of High Capitalism*, 2nd ed. (London, 1997), Susan Buck-Morss, 'The Flâneur, the Sandwichman and the Whore: The Politics of Loitering', *New German Critique* 39 (1986), Mary Gluck, 'The Flâneur and the Aesthetic: Appropriation of Urban Culture in Mid-19th-Century Paris', *Theory, Culture & Society* 20(5) (2003), Keith Tester (ed.), *The Flâneur* (London, 1994), Elizabeth Wilson, 'The Invisible Flaneur', *New Left Review* 191 (1992).

17    Roman Jakobson and Claude Lévi-Strauss, '"Les Chats" De Charles Baudelaire', *L'Homme* 2(1) (1962). One critic has noted drily that 'interpretation of the "Cats" has become a sub-category of Baudelaire criticism, very rich in summing up several tendencies

concerning the relationship between Baudelaire's cats and the relationship that he drew between life, experience and truth in modernity. In his prose poem 'The Clock', Baudelaire recounts the story of a missionary in China, who after asking a little boy for the time, was surprised to see the boy return with a very fat cat, look in the whites of its eyes, and announce (correctly) that the time was not quite noon. 'The Chinese', he writes, 'can tell the time in the eyes of cats'.[18] Then the narrator looks in the eyes of his beautiful Féline – it is ambiguous whether Féline is a cat or his human lover – and finds that 'in the depths of her adorable eyes I always see the hour distinctly, always the same hour, an hour vast, solemn, and grand as space, without divisions into minutes and seconds – a motionless hour unmarked by clocks, but light as a sigh, rapid as the blink of an eye'.[19] The eyes of the cat – or the eyes of a woman looked upon *as* a cat – offer a window onto a time that transcends clock time, a form of time that has the absolute breadth of space itself. Through the cat's eyes, a fleeting glimpse of an other-worldly mode of temporal existence – a fragment of the Ideal, an eternal truth beyond everyday appearance – is gained.

For some writers, the human gaze is the very foundation of society.

> By the glance which reveals the other, one discloses himself. By the same act in which the observer seeks to know the observed, he surrenders himself to be understood by the observer. The eye cannot take unless at the same time it gives. The eye of a person discloses his own soul when he seeks to uncover that of another.[20]

The exchange of eyes in this poem, however, is not between human eyes, but between human and non-human. The narrator surrenders himself to an inhuman gaze, and finds in it the Ideal, a truth beyond appearance. In doing so, he makes a surprising acknowledgement: the animal can *respond* to the human. In the sixteenth century, Western thought created a rigid epistemological barrier between the human and the nonhuman. Only humans, it asserted, have a relationship with

---

of the last twenty years of literary analysis, but unnecessarily complex for the reader who is interested in Baudelaire'. Graham Robb, '"Les Chats" De Baudelaire: Une Nouvelle Lecture', *Revue d'Histoire littéraire de la France* 6 (1985). 'L'exégèse des «Chats» est devenue un sous-genre de la critique baudelairienne, très riche en ce qu'il résume plusieurs tendances de l'analyse littéraire des vingt dernières années, inutilement complexe pour le lecteur qui s'intéresse à Baudelaire'.

18   Charles Baudelaire, *Paris Spleen and La Fanfarlo*, trans. Raymond MacKenzie (Indianapolis, 2008), p. 30.

19   Ibid.

20   Georg Simmel, 'Sociology of the Senses: Visual Interaction', in Robert Park and Ernest Burgess (eds), *Introduction to the Science of Sociology* (Chicago, 1924), p. 358. See, for example, Martin Jay, *Downcast Eyes: The Denigration of Vision in Twentieth-Century French Thought* (Berkeley, CA, 1993), Anthony Synnott, 'The Eye and I: A Sociology of Sight', *International Journal of Politics, Culture, and Society* 5(4) (1992).

truth and knowledge; and only humans possess the active capacity to respond. In this sense, the animal is wholly 'other' to mankind. But in this poem the narrator submits to this otherness and finds in it an experience of truth that is more powerful than anything he can gain from the world of human affairs. In the gaze of a cat, he encounters the possibility of an exterior truth, previously unknown, troubling the limits of human life. As Jacques Derrida puts it,

> As with every bottomless gaze, as with the eyes of the other, the gaze called animal offers to my sight the abyssal limit of the human: the inhuman or the ahuman, the ends of man, that is to say the bordercrossing from which vantage man dares to announce himself to himself, thereby calling himself by the name that he believes he gives himself.[21]

This prose poem, then, opens up the possibility of a gaze that troubles the limit between human and animal, and finds in this transgression the source of a new experience of the Ideal and a novel encounter with truth. Yet, as Baudelaire explores in another prose poem 'Laquelle est la Vrai?' (Which is the True One?), such glimpses of the Ideal, in modernity, can only be gained from the vantage point of a truth that is grubby, prosaic and common. In this story a young woman named Benedicta, whom he recently met and 'who radiated the ideal', dies suddenly ('too beautiful to live very long'), and it is left to the narrator to bury her.[22] Immediately afterwards he spies a little person, looking just like the woman he buried, stamping on the ground with hysterical violence and laughter. 'It's me, the real Benedicta!', she shouts coarsely. 'It's me, a celebrated slut! And as a punishment for your madness and your blindness, you will love me just as I am!' But the furious narrator refuses to love this degraded version of the woman, and stamping his foot on the floor over the grave, finds himself sinking down to his knees into the grave. Now, he concludes, 'I remain, perhaps forever, attached to the grave of the ideal'.[23] The narrator, unable to love this desanctified woman, remains tethered to an ideal that cannot live long in modernity. He cannot accept that truth is not idealized, pristine beauty but something ugly, transitory, and deformed.

Exploring this pathway to poetic truth – one that passes through the fleshy and ugly details of everyday life rather than fleeing them to the realm of the Ideal – was a preoccupation of Baudelaire's. It is clear that he experienced the alienation of life in the modern city very intensely. His great poetic success, however, emerged from the way in which he deliberately exposed himself to the most alienating effects of the metropolis so as to use his own life as an artistic material upon

---

21    Jacques Derrida, 'The Animal That Therefore I Am (More to Follow)', *Critical Inquiry* 28(2) (2002), p. 381.

22    'Benedicta' dervies from 'bene' (well) and 'dico' (speak), thus meaning spoken well or truly.

23    Baudelaire, *Paris Spleen and La Fanfarlo*, p. 80.

which to imprint the emerging physiognomy of modernity.[24] Baudelaire famously described some of the roles which the hero of modernity must adopt: he must be once artist, dandy, flâneur, conspirator, and apache.[25] Much critical debate has gone into deciding which of these postures was most important.[26] Perhaps the most convincing argument, however, is Walter Benjamin's hypothesis that the attitudes represented by the dandy, apache, ragpicker and flâneur were really disguises: they were roles that the poet adopted in the service of his poetic labour. 'Flâneur, apache, dandy, and ragpicker were so many roles to him. For the modern hero is no hero; he is a portrayer of heroes'.[27] Yet these roles were still of the utmost importance, for they were the devices by which Baudelaire could expose himself to the fullest extremities of modern experience. The modern artist, he believed, must live that modernity and let its truth imprint itself upon his being. Thus true heroism, for Baudelaire, was the ability to live at the heart of the unreal, of appearance. Flâneur, dandy, apache and ragpicker are roles through which the poet can surrender to the shocks of modern life. As Baudelaire puts it, the hero of modernity encounters the city as 'an immense reservoir of electrical energy', both shocking and energizing. Exposing himself to the greatest extent possible to the destructive, alienating shocks of modernity is the precondition for exposing the true – destructive – face of modernity in poetic images. Only by exposing himself to modernity's violence and alienation can the poet grasp the destructiveness of life's 'luminous explosion in space' and recover the ideal and the eternal from the heart of the transitory and the ugly.

Baudelaire seems to have found the cat a particularly apt image for this unsettling affinity between poetic truth and the everyday world of appearance, sensation, death and intoxication. In particular, he associated cats with the experience of intoxication and sensory abundance that was central to his view on urban modernity. In 'Le Chat', for example, the cat appears as a double, evoking both the eternal beauty of the Ideal and also the fleeting intoxication of the everyday.

---

24    For a detailed exploration of this point, see Julian Brigstocke, 'Artistic Parrhesia and the Genealogy of Ethics in Foucault and Benjamin', *Theory, Culture & Society* 30(1) (2013).

25    Charles Baudelaire, 'The Painter of Modern Life', in Jonathan Mayne (ed.), *The Painter of Modern Life, and Other Essays* (London, 1964).

26    Jean-Paul Sartre, for example, insists that Baudelaire's most highly valued form of heroism was dandyism – a narcissistic retreat into the security of the isolated self, a celebration of 'the parasite of parasites – the dandy who was the parasite of a class of oppressors'. Jean-Paul Sartre, *Baudelaire* (New York, 1967), p. 146. Michel Foucault, by contrast, also places emphasis on Baudelaire's dandyism, but celebrates this as an ethical commitment to the recreation of subjectivity, rather than a disavowal of ethical duty. See Michel Foucault, 'What Is Enlightenment?', in Paul Rabinow (ed.), *Ethics, Subjectivity and Truth: The Essential Works of Foucault 1954–1984, Volume One* (London, 2000).

27    Walter Benjamin, 'The Paris of the Second Empire in Baudelaire', in Howard Eiland and Michael Jennings (eds), *Walter Benjamin: Selected Writings, Volume Four, 1938–1940* (Cambridge, MA, 2006), p. 60.

Come, my fine cat, to my amorous heart;
Please let your claws be concealed.
And let me plunge into your beautiful eyes,
Coalescence of agate and steel.

When my leisurely fingers are stroking your head
And your body's elasticity,
And my hand becomes drunk with the pleasure it finds
In the feel of electricity,

My woman comes into my mind. Her regard
Like your own, my agreeable beast,
Is deep and is cold, and it splits like a spear,

And, from her head to her feet,
A subtle and dangerous air of perfume
Floats always around her brown skin.[28]

The poem directly evokes the sensual qualities of nature in modernity. The cat, an animal incarnation of the narrator's lover, is erotically charged, a bestial metaphor for the narrator's partner. It excites all of the senses, matching sensual exuberance with a disquieting sense of danger. It is here that its double nature is clearest. Its gaze is magnetic, irresistible ('Let me plunge into your beautiful eyes') but also violent and piercing: its regard 'is deep and is cold, and it splits like a spear'. To the sense of touch it offers invigorating intoxication: when the narrator strokes its head, his hand 'becomes drunk with the pleasure it finds / In the feel of electricity'. This is only after, however, he has begged the cat to 'please let your claws be concealed'. Its smell is gentle and calming; yet this too is accompanied by a palpable sense of peril: 'a subtle and dangerous air of perfume / Floats always around her brown skin'.

In another poem, also called 'Le Chat', however, it is *sound* that most thoroughly penetrates the narrator's soul:

---

28    Charles Baudelaire, *The Flowers of Evil*, trans. James McGowan (Oxford, 1993), pp. 70–71.

    'Viens, mon beau chat, sur mon cœur amoureux;
Retiens les griffes de ta patte,
Et laisse-moi plonger dans tes beaux yeux,
Mêlés de métal et d'agate.
Lorsque mes doigts caressent à loisir
Ta tête et ton dos élastique,
Et que ma main s'enivre du plaisir
De palper ton corps électrique,
Je vois ma femme en esprit. Son regard,
Comme le tien, aimable bête
Profond et froid, coupe et fend comme un dard,
Et, des pieds jusques à la tête,
Un air subtil, un dangereux parfum
Nagent autour de son corps brun'.

...
When he meows, one scarcely hears,

So tender and discreet his tone;
But whether he should growl or purr
His voice is always rich and deep.
That is the secret of his charm.

This purling voice that filters down
Into my darkest depths of soul
Fulfils me like a balanced verse,
Delights me as a potion would.

It puts to sleep the cruellest ills
And keeps a rein on ecstasies –
Without the need for any words
It can pronounce the longest phrase.

Oh no, there is no bow that draws
Across my heart, fine instrument,
And makes to sing so royally
The strongest and the purest chord,

More than your voice, mysterious cat.[29]

Here, as with Poe's tale, the cat evokes a sensory experience that is beyond words and representation – a penetrating, direct music that pierces the depths of the human soul.

---

29    Ibid., 102–5.
'Quand il miaule, on l'entend à peine,
Tant son timbre est tendre et discret;
Mais que sa voix s'apaise ou gronde,
Elle est toujours riche et profonde.
C'est là son charme et son secret.
Cette voix, qui perle et qui filter
Dans mon fonds le plus ténébreux,
Me remplit comme un vers nombreux
Et me réjouit comme un philtre.
Elle endort les plus cruels maux
Et contient toutes les extases;
Pour dire les plus longues phrases,
Elle n'a pas besoin de mots.
Non, il n'est pas d'archet qui morde
Sur mon coeur, parfait instrument,
Et fasse plus royalement
Chanter sa plus vibrante corde,
Que ta voix, chat mystérieux,
Chat séraphique, chat étrange,
En qui tout est, comme en un ange,
Aussi subtil qu'harmonieux!'

The figure of the cat stages the sensual hedonism, but also the existential danger, of the five senses. Beauty, for Baudelaire, always has something strange in it, something ugly, something dangerous. The cat is a reincarnation of erotic experience, but it also testifies to the recognition that genuine experience is an experience of the 'beyond', an experience that tests the limits of experience, the limits of life. Sensory extravagance is haunted by a mysterious gulf that lies inaccessible to organic perception. This is clearest in the poem 'Les Chats', in which cats are like:

> Great sphinxes in the desert solitudes,
> Who seem to be entranced by endless dreams;
> Within their potent loins are magic sparks,
> And flakes of gold, fine sand, are vaguely seen
> Beyond their mystic eyes, gleaming like stars.[30]

Here the sensuality of the cat moves beyond the five senses altogether, towards the mysticism of the Moon – a nothingness that is impenetrable and secretive, seeking 'the silent horror of the night'. The cat doesn't just stage sensory excess, then, but also a blank vacancy within the sensual abundance of modernity, a void that possesses its own life, mystery, and creativity.

Like Poe's tale, Baudelaire's poems use the cat as a way of grasping a distinctively modern form of experience, one that is contradictory, ungraspable and unrepresentable – beyond human reason and human experience. The cat, at once commonplace and mysterious, testifies to the mysteries and beauties that can be found within modernity's mundane, ugly, and distorted bodies and spaces. In the figure of the cat, something of the road towards poetic truth in modernity can be glimpsed – a circuitous route that wanders the streets of the city alone in the dark of night, seeking transitory loves and exploring the furthest extremities of alienated urban living.

## Manet's *Olympia*

Two pairs of eyes meet the viewer's gaze (Figure 2.3). One belongs to a prostitute, naked apart from some jewellery, a flower in her hair, and small bow around her neck – a present shortly to be unwrapped. Although she is thin and diminutive, her gaze is defiant, cold and strong, confronting the viewer with a sexuality that is unadorned and undisguised. The first finger of her left hand disappears, scandalously, from view.

---

30  Ibid., 134–5.
   'Ils prennent en songeant les nobles attitudes
   Des grands sphinx allongés au fond des solitudes,
   Qui semblent s'endormir dans un rêve sans fin;
   Leurs reins féconds sont pleins d'étincelles magiques,
   Et des parcelles d'or, ainsi qu'un sable fin,
   Etoilent vaguement leurs prunelles mystiques'.

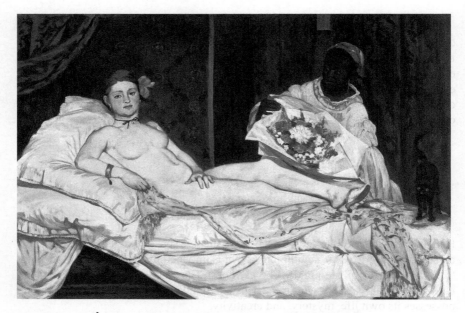

**Figure 2.3    Édouard Manet, *L'Olympia*, 1863. Musée d'Orsay.**
**© RMN-Grand Palais/Patrice Schmidt/Musée d'Orsay, Paris**

The second pair of eyes, contrasting strongly to the first, belongs to the startled black cat at the foot of the bed (Figure 2.4). Its tail erect, its back arched, the cat is erotically charged, standing in for what her left hand refuses to the viewer's gaze.[31] Its yellow eyes are hostile, malevolent. Do these eyes, perhaps, indicate something of what lies behind the inexpressive gaze of Olympia herself, an inhuman window into her hidden soul?

This double gaze is highly unusual. In fact, *Olympia* is Manet's only large multifigure painting in which more than one figure meets the viewer's gaze.[32] One critic speculates that the aim of Manet's single gazes may have been to diminish the possibility of establishing any kind of psychological relationship between viewer and figure. In place of a psychological relationship, Manet creates an effect where 'it is as though the *painting itself* looks or gazes or stares at one – it is as though it confronts, fixes, even *freezes* one … and as though this was an essential source of Manet's conviction, insofar as he achieved that conviction, that the paintings in

---

31    The French word 'chatte' shares the innuendo of the English word 'pussy'. Thus Manet came as close as he dared to breaking the taboo on representing pubic hair in art.

32    Michael Fried, *Manet's Modernism, or, the Face of Painting in the 1860s* (Chicago, 1996), p. 469, note 26.

**Figure 2.4** **Édouard Manet, *L'Olympia*, 1863 (detail). Musée d'Orsay.**
**© RMN-Grand Palais/Patrice Schmidt/Musée d'Orsay, Paris**

question really *were paintings*'.[33] In *Olympia*, however, the double gaze does nothing to diminish the awareness of the viewer of the material qualities of the painting. In fact, the opposite seems to be true. It is this painting, after all, which has been described as the seminal work of modern art, the inauguration of a 'fundamental change soon to come over all European painting' where art was no longer held in representational service, but found its own autonomy, a 'painting that should rise in utter freedom, in natural silence, painting for its own sake, a song for the eyes of interwoven forms and colours'.[34] In this painting, writes Georges Bataille,

> the picture obliterates the text, and the meaning of the picture is not in the text behind it but in the obliteration of that text ... In her provocative literalness she is nothing. Her real nudity (not merely that of her body) is the silence that emanates from her, like that from a sunken ship. All we have is the 'sacred horror' of her presence – presence whose sheer simplicity is tantamount to absence.[35]

---

33   Ibid.
34   Georges Bataille, *Manet*, trans. Austryn Wainhouse and James Emmons (London & Basingstoke, 1983), p. 35.
35   Ibid., 62.

The painting can usefully be viewed as an exploration of the power of negation – so that the picture's success lies in the way in which it makes use of presence to destroy itself, leaving only an absence, a silent horror. Olympia is a prostitute, completely undisguised and unadorned. Described by contemporaries as gorilla-like and corpse-like, she is devoid of any idealized beauty, existing only as a forceful assertion of erotic presence. The Second Empire was the great age of the courtesan, and in *Olympia* this sexual economy was laid bare in art for the first time.[36] The realism of the canvas confronted the bourgeois male viewer – and, perhaps, the wife and children who accompanied him to the exhibition – with the uncomfortable truth of his complicity in this unmentionable but thriving sexual economy. The figure was 'obviously naked rather than conventionally nude', and 'her wide eyes, imperturbable expression, and impertinent attitude seemed ... to force the spectator to assume that he was in the same room with her. No conventional generalizations of face or figure mitigated the disconcerting surprise of seeing a modern young woman in a situation few ladies and gentlemen could publicly acknowledge'.[37]

The painting caused one of the great cultural scandals of the nineteenth century, and the black cat at the foot of Olympia's bed, along with the black servant at her side, encapsulated much of what Manet's contemporaries believed to be scandalous and inappropriate.[38] The negress conveyed a primitive, exotic sensuality: since the eighteenth century, black women had been widely regarded as being more uninhibited and passionate than white women.[39] Similarly, the cat evoked promiscuity and moral indifference. The exaggerated length and rigidity of the cat's tail had obvious phallic significance. Moreover, cats were associated with rapacious female sexuality. In his *Histoire Naturelle*, in comments repeated in the 1867 Larousse *Grand Dictionnaire Universel du XIX Siècle*, the biologist Buffon had suggested that the female cat was more driven by desire than the male. The female cat has an overwhelming need for sex: 'she invites it, calls for it, announces her desires by her piercing cries, or rather, the excess of her needs'. The female forces herself on the reluctant male, 'and when the male runs away from her, she pursues him, bites him, and forces him, as it were, to satisfy her'.[40] Moreover, the cat was a *wild* animal, never susceptible to domestication in the manner of dogs.

---

36    On prostitution in Second Empire France, see Charles Bernheimer, *Figures of Ill Repute: Representing Prostitution in Nineteenth-Century France* (Cambridge, MA, 1989), Alain Corbin, *Women for Hire: Prostitution and Sexuality in France after 1850* (Cambridge, MA, 1990), Joanna Richardson, *The Courtesans: The Demi-Monde in Nineteenth-Century France* (London, 1967), Virginia Rounding, *Grandes Horizontales: The Lives and Legends of Four Nineteenth-Century Courtesans* (London, 2003).

37    George Heard Hamilton, *Manet and His Critics* (New Haven, 1954), pp. 67–8.

38    See Theodore Reff, *Manet: Olympia* (London, 1976), pp. 89–101.

39    Ibid., 92.

40    Georges Buffon, *Histoire Naturelle, Générale et Particulière* (Paris, 1769). Repeated in Pierre Larousse, 'Chat', *Grand Dictionnaire Universel du XIX Siècle, Tome Troisième* (Paris, 1867). This and the quotes that follow it in this paragraph are cited in Kete, *The Beast in the Boudoir: Petkeeping in Nineteenth-Century Paris*, pp. 118–19.

Cats have 'an innate malice, a falseness of character, a perverse nature, which age augments and education can only mask'. They are deceitful and untrustworthy: 'They take easily to the habits of society, but never to its moral attitudes; they only appear to be affectionate'. They are deceitful: 'either out of mistrust or duplicity, they approach us circuitously when looking for caresses that they appreciate only for the pleasure they get from them'. Indeed, Alphonse Toussenel, in his 1855 *Zoologie Passionnelle*, explicitly compared the cat to the prostitute.

> An animal so keen on maintaining her appearance, so silky, so shiny, so eager for caresses, so ardent and responsive, so graceful and supple ... an animal who makes the night her day, and who shocks decent people with the noise of her orgies, can have only one single analogy in this world, and that analogy is of the feminine kind.[41]

Together, the negress and the cat contributed to what Paul Valéry referred to as *Olympia*'s 'primitive barbarity and ritual animality'.[42]

The figure of the black cat added to the scandalous nature of the painting through its reference to unadorned sexuality and bestial promiscuity. However, it also plays a more formal role in the composition of the painting. One of the striking features of *Olympia* is the tonal range employed: Manet employs a close range of light shades and a close range of dark shades, but with very subdued intermediate tones. The viewer has to make an effort to distinguish the closely related shades of light and dark. Thus the painting makes use of both a bold contrast between white and black, but also fine distinctions within those light and dark shades. Indeed, Emile Zola, defending the painting, justified the black cat and servant on precisely these grounds, arguing that they were included merely to make possible this striking tonal contrast. Even if Zola was merely trying to draw attention from the scandalous aspects of Manet's painting, it remains true that the tonal range of the painting is very powerful, emphasizing the stony cold blankness of Olympia's eyes, and dramatizing a stark opposition between her eyes and those of the cat.

Unlike Olympia's eyes, the cat's stare hides nothing: it conveys blatant hostility. The cat, tensed and with raised back, seems ready to attack the viewer. Since it clearly stands in for Olympia's sexuality, however, it seems that the painting portrays, not so much a real cat's gaze, but rather a negative image of Olympia's own. In light and dark, exteriority and interiority, impassivity and rage, Olympia meets our eyes twice simultaneously. What she conceals with her own gaze, she reveals in the cat's. In this way, the sheer presence of her nakedness is doubled; not only her outward appearance but her inner animal drives, via the cat's gaze, stare at the viewer. Everything becomes opened to view. Olympia becomes a pure presence, her body unfolded before the viewer. But through this series of contrasts and doubles, it becomes clear that the power of the painting comes less from

---

41   Kete, *The Beast in the Boudoir: Petkeeping in Nineteenth-Century Paris*, p. 120.
42   Paul Valéry, 'The Triumph of Manet'.

presence than from the impression of its negative image, absence, that is created. The truth that the painting presents to the viewer is, finally, not simply a truth concerning the mid-nineteenth century sexual economy, but a truth *about* truth. Artistic truth, it shows us, must from henceforth be conveyed through absence rather than presence. Truth will no longer emanate from the interior of the subject, but from what lies *beyond* it. Truth must be discerned, not in the clear light of day, but in the silent horror of a black night.

**Truth and the Cynicism of Culture**

In modernity, writes Michel Foucault, we have forgotten a *spiritual* experience of truth that was at the heart of the ethics of Antiquity. For the Ancient Greeks, the practice of 'parrhesia', or courageous truth telling, was considered an important ethical duty. Parrhesia was a verbal activity in which speakers expressed a personal relationship to truth, and were willing to risk themselves to voice uncomfortable truths to the powerful, because they recognized it as an ethical duty both to themselves and to others. In modernity, however, the experience of truth has become dislocated from this active, personal relationship with truth. It has become something secure, timeless, and passive – something indistinct from knowledge. By contrast, in Antiquity, truth was distinct from knowledge, because an essential part of the experience of truth was the active transformation of the self. Parrhesia was a way 'for the subject to constitute himself ethically at the moment he tells the truth, and to transform himself to become capable of a truthful statement'.[43] In the equivalence it draws between truth and knowledge, modernity has all but forgotten a transformative spiritual experience of truth. '[W]e could call "spirituality"', he writes, 'the search, practice, and experience through which the subject carries out the necessary transformations on himself in order to have access to the truth'.[44] Spirituality postulates that truth is not given to the subject by right, but that the subject must be changed, transformed, something other than itself, to gain right of access to truth.[45] Encountering truth requires a transformation of the subject.

Foucault was particularly interested in the parrhesia of the Cynics, exemplified by Diogenes of Sinope, who sought to provoke through a 'life of scandal'. The Cynics were committed, not just to truth-telling, but to living a truthful life. The Cynic's resolve was to *expose* life, in all senses of the word. Through a series of public demonstrations, from verbal excoriation to extreme asceticism to public masturbation, Diogenes put his life on display and at risk; he sought to hide

---

43    Alexandre Macmillan, 'Michel Foucault's Techniques of the Self and the Christian Politics of Obediance', *Theory, Culture & Society* 28(3) (2011), p. 10.

44    Michel Foucault, *The Hermeneutics of the Subject: Lectures at the Collège de France 1981–1982,* trans. Graham Burchell (Basingstoke and New York, 2005), p. 15.

45    See Paul Rabinow, 'Foucault's Untimely Struggle: Toward a Form of Spirituality', *Theory, Culture & Society* 26(6) (2009).

nothing of his life from view. Foucault speculates that this Cynical experience of truth may not have been destroyed by Christianity. Instead, it travelled across Western culture in the idea that an ethical form of life can be a vehicle for disclosing truth. It did this in religious movements, as well as in the model of life as a violent, scandalous manifestation of truth that was adopted by revolutionary movements over the nineteenth century. Additionally, however, a third medium of Cynicism in European culture can be found in *art*.[46] Citing writers and artists such as Baudelaire, Manet, Beckett and Bacon, Foucault observes how, in modernity, 'art ... establishes a polemical relationship of reduction, refusal, and aggression to culture, social norms, values, and aesthetic canons'.[47] In modern art one sees an

> endless movement by which every rule laid down, deduced, induced, or inferred from preceding actions is rejected and refused by the following action. In every form of art there is a sort of permanent Cynicism towards all established art ... reduction, laying bare the basics of existence; permanent refusal and rejection of every form of established art.[48]

Modern art, Foucault goes on,

> is Cynicism in culture; the Cynicism of culture turned against itself. And if this is not just in art, in the modern world, in our world, it is especially in art that the most intense forms of a truth-telling with the courage to take the risk of offending are concentrated.[49]

The kinds of modern art exemplified by Baudelaire and Manet, then, can be viewed in terms of their distinctive role in the modern ethics of truth: the links they draw between art, truth and subjectivity.

If Poe, Baudelaire and Manet can indeed be understood as contributing to a distinctively modern spiritual experience of truth, then the black cat can be seen to emerge as an important figurative device through which they evoked this truth. In different ways Poe, Baudelaire and Manet each used the cat in order to confront the modern subject with a bestial form of subjectivity, one that eluded the sanctified limits of the human subject. The cat, for each of these artists, was useful as a way of representing a truth based in the experience of absence and darkness; a truth that lurked in deep shadows, dangerous affections, and muddy streets. Through the figure of the black cat, each of these artists could evoke an experience of truth

---

46   Michel Foucault, *The Courage of Truth: The Government of Self and Others, Volume Two. Lectures at the Collège de France, 1983–1984* (Basingstoke, 2011), pp. 185–6. See Joseph Tanke, *Foucault's Philosophy of Art: A Genealogy of Modernity* (London, 2009).

47   Foucault, *The Courage of Truth: The Government of Self and Others, Volume Two. Lectures at the Collège de France, 1983–1984*, p. 188.

48   Ibid.

49   Ibid., 189.

that tested the limits of the human subject, evoking animalistic and supernatural forces that threatened to confront the subject with something wholly other to itself. The black cat exemplified a kind of 'limit-experience', an experience at the extremities of reason, sexuality and organic life itself. Many of these iconographic connotations of the black cat, as we will see, were appropriated by the popular culture of fin-de-siècle Montmartre.

I have suggested that art plays an important role in speaking the truth, via the outside of life, in the modern metropolis. Effectively speaking the truth, however, requires finding ways of making artistic speech acts *authoritative*, of lending them weight and power in the urban flow of words and ideas. Late nineteenth-century France saw a proliferation of discourses concerning the crisis of authority that accompanied the waning of tradition, religion and aristocratic rule. For Montmartre bohemians, this crisis opened up a powerful new challenge: might it be possible for the arts to capitalize on this crisis of authority to harness new, distinctively modern and creative forms of social and political authority?

# Chapter 3

# Order and Progress:
# Crises of Authority and the Pursuit of
# Experience in the Early Third Republic

## Introduction

The birth of the Third Republic marked the increasing dominance of an art of governing that drew its legitimacy from the power of science and reason to make societal life perceptible and transformable. The biopolitical culture of fin-de-siècle Montmartre, we will see, attempted to ground a new authority to make social and political claims on behalf of the poor and marginalized by testing and contesting the accepted frameworks for experiencing, representing and performing the life of the city. This form of counter-cultural, 'immanent' authority was based on direct experience rather than scientific, political or religious representation. First, however, we must contextualize this argument within wider fin-de-siècle debates concerning the 'crisis of authority' in the Third Republic. In this chapter, I analyze political debates in 1870s and 1880s on this crisis of authority in order to build a novel theoretical model of three competing modes of authority production in fin-de-siècle France: 'transcendent' authority; 'experimental' authority; and 'experiential' authority. The artistic and political experiments in fin-de-siècle Montmartre attempted to capitalize on a radically new form of authority in order to effectively speak the truth to power: this was the authority of *experience*: its dynamism, its creativity, and its links to the life of the body and the vitality of the city.

By the 1880s, the bitter contest between a past-oriented, conservative emphasis on traditional values such as family and community as sources of vitality, and a future-oriented, republican emphasis on reason and experimental observation as a way of fostering life scientifically, had been definitively settled (at least with respect to the nation's constitutional and political structure). Politically, France had emerged as a representative democracy with universal male suffrage (the first nation in Europe to achieve this), with parliamentary chambers filled with doctors and lawyers. However, as we will see, there was an emerging third source of authority under construction: an experiential authority that can be distinguished from both the conservative emphasis on morals and the republican emphasis on scientific observation. Whilst such experiential authority frequently served hegemonic bourgeois aims, it also offered a tool that could be re-appropriated by those – such as the artists and anarchists of Montmartre – who wished to voice their opposition to the dominant regime.

## Order and Progress

In the 1870s, fresh from military defeat, invasion, and civil unrest, political theorists argued that the most important task France faced was to reverse national decline and kick-start the march of progress. But this required, first and foremost, imposing order. 'Society', wrote Emile Littré, one of the most influential political theorists of the early Third Republic, 'is governed by two equally powerful tendencies: order and progress'.[1] A number of related social and political crises, however, seemed to pose a serious threat to any attempt to re-impose order upon French society.

Perhaps the most important crisis was one of *authority*. The new Republic had few resources with which to ground its authority and its claim to obedience. The Commune had challenged the legitimacy of its rule, and the state, unable to exert the authority to command free obedience, found itself compelled to force the issue through the use of violence. In opposition to a highly centralized republican state dominated by the votes of the countryside, the fundamental goal of the Commune had been municipal autonomy, aiming at a radical transformation of the political institutions that represented local society. It had asserted 'a form of popular democracy, articulating grassroots democracy and representative democracy to reorganize the nation by the connection between successive levels of political delegation'.[2] Whilst the Commune itself had been defeated, both its threat to the Republic and the violence with which it had been suppressed prompted concern about the diminishing efficacy of recognized authority structures. The key challenge for the new Republic was to create new grounds upon which to draw a durable, non-traditional, modern form of authority.

This crisis of authority also manifested itself as a *temporal* crisis. The positivist spirit of the 1850s and 1860s, supremely self-confident of a scientific march towards future peace, prosperity and social harmony, had been severely damaged by the perceived irrationalism and regressiveness of the war and its aftermath. On the one hand, all links to the past seemed to have been destroyed: both town and countryside had been transformed by urbanization and industrialization, and a new spirit of secularism and republicanism seemed to be replacing the old pillars of altar and crown. Everything seemed to be in continual flux. On the other hand, the path towards the future also seemed to have been broken. The nation, defeated and demoralized, seemed to be regressing into the past rather than marching confidently into the future. The future threatened further degeneration rather than inevitable evolution.

---

1    Emile Littré, *Application de la Philosophie Positive au Gouvernement des Societes et en Particulier a la Crise Actuelle* (Paris, 1850), p. 34.

2    M. Castells, 'Cities and Revolution: The Commune of Paris, 1871', *The City and the Grassroots: A Cross-Cultural Theory of Urban Social Movements* (Berkeley, 1983), p. 25.

The present, its links to both the past and the future irreparably damaged, seemed in danger of withering away.

Finally, the nation faced a serious *urban* crisis.[3] The Commune had been, more than a class struggle, an urban revolution. As Manuel Castells highlights, the most important demand of the Commune – the first to be implemented – was the cancellation of rents and the attempt to curb the forms of speculation that were leading to ever higher rents.[4] The revolt threw into relief an appalling housing crisis. Urbanization had brought hundreds of thousands of provincial immigrants to the city, many of them finding work building the new dwellings that they could never hope to live in. The grandiose rebuilding of Paris had caused popular neighbourhoods of Paris to gentrify or disappear, meaning that those displaced from the centre joined those coming from the countryside to overcrowd the remaining popular areas: Montmartre in the north, Belleville in the east, and the Butte-aux-Cailles in the south east. Living conditions were squalid, rents were very high, and tenants were immediately evicted upon late payment of rents. The revolt – which came from precisely those areas where the housing problem was most acute – seemed, to both the monarchist right and the republican left, confirmation of the hypothesis that the urban environment was proving deleterious to public health and public morality, and in doing so, playing a key role in the collapse of authority and temporality.

One thing, then, was generally agreed upon. There was a pressing need for a regeneration of the social body. The population was disturbingly lacking in vigour, vitality and strength, and this weakness was leading them to depravity that would multiply as the population grew.[5] 'Society is ill', wrote Gustave Orlay, 'and those who have the ambition to be its doctors must make themselves perfectly aware of its intimate constitution, in fear of applying remedies to it that are worse than the evil'.[6] This metaphor of illness expressed key aspects of the political crisis: the crisis of authority (the need for a wise leader to heal the social body); the crisis of temporality (the possibility of making society regress still further rather than healing it); and the crisis of the city (the problem of unhealthy living spaces making the population morally and physically diseased).

It is possible to group the responses to these crises into three categories. The first was a conservative and nostalgic appeal to *transcendent* authority: to the pre-revolutionary values of God, tradition, and community. The second and third models, by contrast, both involved appeals to an *immanent* form of authority.

---

3   See David Harvey, *Paris: Capital of Modernity* (New York, 2003).

4   Castells, 'Cities and Revolution: The Commune of Paris, 1871'.

5   Robert Nye, *Crime, Madness & Politics in Modern France: The Medical Concept of National Decline* (Princeton, 1984).

6   Gustave Orlay, *L'Autorité & La Liberté Opposées au Despotisme du Nombre* (Paris, 1874). 'La Société est malade, et ceux qui ont la prétention d'être ses médecins doivent se rendre parfaitement compte de sa constitution intime, dans la crainte de lui appliquer des remèdes pires que le mal'.

Enlightenment, Kant had argued, involves freeing oneself from the dictates of external (transcendent) authority.[7] Instead, he had suggested, one must learn to think autonomously, to mobilize the authority of reason as something immanent to individual experience. One attempt to do so, in the spirit of this Kantian model of Enlightenment, was the positivist attempt to ground societal progress on the authority of experimentation, and hence, the expertise of scientific representation. Another model, however – one that will be explored in detail throughout this book – was the effort to harness the authority of *corporeal experience* to support radical truth claims against dominant powers of the Third Republic.

## Transcendent Authority

In republican thought, authority was a concept strongly associated with the political right – in particular the Legitimists who hoped for a restoration of the monarchy and the coronation of the Bourbon heir the Comte de Chambord.[8] In the years following the Commune, authority was associated with Patrice MacMahon's monarchist and pro-Catholic government, and its notorious attempts to engineer a return to 'moral order'.[9] This was a form of order based on individual morality, and hence a demonization of working class 'vices' such as alcoholism, vagrancy and prostitution. The individual, it was thought, 'needed authority to guide him, roots to give him a sense of identity, and strong social institutions like the family and the Church to integrate his life with that of the wider community'.[10] Monarchism was closely allied with Catholicism, which celebrated the principle of sovereign authority as represented by King and Pope.

For the political right, the Commune confirmed their sense of the disastrous consequences of the erosion of authority and community through the rapid urbanization that was caused, they argued, by peasants' greed (that is, their exodus to cities in search of higher wages). Conservatives refused to recognize the legitimacy or effectiveness of any form of authority not based on tradition and community. The decline in authority, they insisted, was directly responsible for revolutionary violence, crime, deviance, and degeneration. In opposition to the individualism of the

---

7    Immanuel Kant, 'What Is Enlightenment?', in L. Beck (ed.), *Selections* (New York, 1988).

8    Peter Davies, *The Extreme Right in France, 1789 to the Present: From De Maistre to Le Pen* (London, 2002), Samuel M. Osgood, *French Royalism since 1870*, 2nd ed. (The Hague, 1970), R. Rémond, *The Right Wing in France. From 1815 to De Gaulle.*, trans. James Michael Laux (Philadelphia, 1966), pp. 55–78, Martin Simpson, 'The Death of Henri V: Legitimists without the Bourbons', *French History* 15 (2001).

9    Robert R. Locke, *French Legitimists and the Politics of Moral Order in the Early Third Republic* (1974).

10    Robert David Anderson, *France 1870–191: Politics and Society* (London, 1977), p. 103.

republican left, conservatives viewed social life in terms of family and community. They saw individual life to be meaningless if not integrated into a developed family life, since the essence of an individual's life came from membership of this group whose bonds were sealed by the inheritance of blood through the generations.[11] They were fiercely anti-statist, wanting instead to preserve or restore forms of community in which social functions were performed by nobles and priests, rather than state bureaucrats. 'The state's claim to a life of its own violated their conception of a society in which the father held his position in the family, the priest's in the parish, and the notable's in the community by prescriptive right'.[12]

A number of prominent political theorists wrote books and pamphlets deploring this collapse of traditional authority. Such publications invariably argued that the state should be modelled on the family, regarding the absolute authority of the father over his family as the appropriate model for state power. As Orlay wrote, in a work titled *Authority and Liberty Opposed to the Despotism of Number*: 'The family is the base and image of Society. What are the fundamental principles of the family, the laws that are essential to its existence? They are: 1. The uncontested authority of the father exercised in the common interest; 2. Mutual respect between families for justly acquired rights to property, the traditional result of work and saving. Thus in the State, one needs a respected authority that is exerted in the interest of all indiscriminately, and which safeguards the general interest, the family, property, dignity, and individual freedom'. 'God', he went on, 'produced man to live in a family, in society, and neither the family nor society are possible without Authority'.[13] The only way of ensuring a return to moral order was to overturn the despotism of the masses: 'Let us return to Order by Authority, and, so that the possession of Authority does not become the occasion and the reason for fights, of competitions that are dangerous for our safety and our peace, let us put it above the covetousness of the multitude: let us return to Monarchy'.[14]

11   Locke, *French Legitimists and the Politics of Moral Order in the Early Third Republic*, p. 144.

12   Ibid., 156.

13   Orlay, *L'Autorité & La Liberté Opposées au Despotisme du Nombre*, pp. 16, 21. 'La famille est donc la base et l'image de la Société: Quels sont les principes primordiaux de la famille, les lois essentielles à son existence? Ce sont: 1° l'autorité incontestée du père s'exerçant dans l'intérêt commun; 2° le mutuel respect entre familles des droits justement acquis, de la propriété, résultat traditionnel du travail et de l'épargne. Ainsi dans l'Etat, il faut une autorité respectée s'exerçant dans l'intérêt de tous indistinctement, sauvegarde de l'intérêt général, de la famille, de la propriété, de la dignité et de la liberté individuelles ... Dieu a produit l'homme pour vivre en famille, en société, et que ni la famille ni la société ne sont possibles sans l'Autorité'.

14   Ibid., 34. 'Revenons à l'Ordre par l'Autorité, et, pour que la possession de l'Autorité ne devienne pas l'occasion et le motif de luttes, de compétitions dangereuses pour notre sécurité, notre tranquillité, mettons-la au-dessus des convoitises de la multitude: revenons à la Monarchie'.

The sanctity of this link between the authority of God, King and father was asserted over and over again. 'God, by his nature', wrote Félix de Marc, 'is infinite power, without limits, without rival. He is the creator of all things in the moral order and the material order. He is not only the source, but also, by virtue of his limitless power, the absolute Master of all that he produced, of everything that exists, of any authority, any justice, any law, any force. All that comes from him, can only belong only to him, otherwise it would restrict his power. Authority, like everything, emanates from him, belongs to him, in an exclusive way; it can exist from the power that he has in him. The first man was born under this sole divine power, his natural and positive law'.[15]

The decline of such forms of transcendent authority is the subject of a celebrated essay by Hannah Arendt. In her essay 'What is Authority', Arendt traces a genealogy of authority in European political philosophy, finding its origins of Ancient Rome and its conviction of the sacredness of foundations (in particular, the founding of Rome).[16] Arendt draws out the etymological link between *auctoritas* (authority) and *augere* (to augment) to argue that traditional authority is based on the augmentation of foundations: not just preserving them, but building on them. Through tradition, every moment adds to the whole weight of the past. Thus authority figures have to possess the quality of *gravitas,* the ability to bear this weight. The growth of society, under this model of authority, is directed to the past rather than the future.

It was such a form of authority, one bound to tradition and religion, that conservatives were still trying to preserve in the 1870s and 1880s. Advocates for the restoration of order through authority grounded their claims upon the stable foundations of immutable truths – truths that were not to be discovered through science but, rather, revealed by God through his appointed intermediaries. The authority of the father, priest and notable can be argued to have been based, first, on an augmentation of communal experience through tradition and inheritance – an accumulation of wisdom, knowledge and experience through time. Second, their authority was based on demonstrating personal gravitas, making visible the great weight of tradition bearing down on them, and their ability to bear this weight and to use it for the guidance of others. Finally, their authority was

---

15  Félix de Marc, Légitimité et la Révolution: Étude sur le Principe D'Autorité (Paris, 1882), p. 103. 'Dieu, par sa nature, est la puissance infinie, sans bornes, sans rivale. Il est le créateur de toutes choses dans l'ordre moral et dans l'ordre matériel. Il est non seulement la source, mais encore, en vertu de ce pouvoir sans limite, le maître absolu de tout ce qu'il a produit, de-tout ce qui existe, de toute autorité, de toute justice, de tout droit, de toute force. Tout ce qui est vient de lui, ne peut appartenir qu'à lui seul, à moins de restreindre sa puissance. L'autorité, comme toute chose, émane de lui, lui appartient en propre, d'une manière exclusive; il ne peut exister de pouvoir qu'en lui et par lui. Le premier homme est né sous cette seule puissance divine, sa loi naturelle et positive'.

16  Hannah Arendt, 'What Is Authority?', *Between Past and Future: Eight Exercises in Political Thought* (New York, 1961).

based on displaying their proximity to an outside, transcendent power. Traditional authority figures asserted their proximity to God and their appointed role as God's intermediaries. Thus they wielded authority by demonstrating their close connection to transcendent power.

One focal point of the spatial recuperation of transcendent authority, as David Harvey has described, was the building of the Sacré Coeur Basilica in Montmartre. Built by the highly conservative, monarchist sympathising cult of the Sacred Heart, the Sacré Coeur was intended to expiate the sins of the rebellious city and to stand as a symbol of the return to 'moral order' and traditional authority structures. The basilica was built on the site at which General Thomas and General Lecomte had been shot during the attempted de-arming of the National Guard that triggered the Commune. It was supported by the new Archbishop of Paris, who was filling the shoes of the man who, along with 24 other priests, had been shot by Communards during the 'semaine sanglante' (bloody week) of May 1871, during which over 20,000 Communards were slaughtered. The Basilica aimed to restore Montmartre's earlier reputation as a site of Christian martyrdom. In some ways, it achieved this, as the anarchist movement of the 1890s reappropriated the discourse of martyrdom in marking the execution of celebrated rebels such as Ravachol (see Chapter 8).

The power of such nostalgic longings for traditional forms of authority, however, appeared to be waning quickly. By the 1880s, the threat of a monarchist restoration, which had seemed almost inevitable in the early 1870s (until the legitimist heir the Comte du Chambord turned down the crown he was offered, in protest at the colour of the flag of the nation he was to rule) had passed.[17] It was reference to the future, rather than the past, which legitimated the increasingly hegemonic forms of positivist, *experimental* authority in the Third Republic.

### Experimental Authority

Whereas conservative theorists and politicians attempted to impose order through an authority that would be grounded on transcendent, divine origins, more individualistically minded republicans, inspired by the positivist philosophy of writers such as Auguste Comte and Émile Littré, were keen to find a *scientific* foundation for social and political authority.[18] Positivism aimed to supersede two

---

17    On being offered the crown by the ardently monarchist President of the Republic, Marshal MacMahon, in 1873, the Comte du Chambord refused to rule a country whose flag was the Republican Tricolore, and insisted on the restoration of the monarchist fleur de lis. It was this quaintly aristocratic stubbornness that saved the Republic.

18    On positivism, see Stanislas Aquarone, *The Life and Works of Emile Littré 1801–1881* (Leyden, 1958), Donald Geoffrey Charlton, *Positivist Thought in France During the Second Empire, 1852–1870* (Oxford, 1959), Frederick Copleston, *A History of Philosophy Volume 9: Maine De Biran to Sartre* (1977), Sudhir Hazareesingh, *Intellectual*

earlier stages of human development, a theological stage and a metaphysical stage, with a third, *positive*, stage, in which a science of humanity could successfully be founded on 'an exact view of the real facts of the case'.[19] 'Positive' philosophy aimed to realize organic qualities of precision, certainty, usefulness, reality and relativity. A positivist study of real facts, Comte argued, could 'direct the spiritual reorganization of the civilized world' by redirecting the worship of God towards the worship of humanity itself.[20]

As Emile Littré made clear, the authority of positive knowledge was not acquired through proximity to transcendent powers, but through access to the immanent forces of material matter itself. Positive science, he wrote, knows of nothing beyond matter and the forces *immanent* in matter. Rejection of the authority of God and tradition, however, did not mean rejecting authority itself. Positivism largely accepted the conservative stress on order, political stability, the value of decisive leadership, and the place of religiosity in human experience. However, it also borrowed from republicanism a belief in the essential values of reason and intellectual freedom. For Littré, social authority was not to come from the personal authority of notables and priests, but from the impersonal authority of scientific expertise. 'The modern advance of civilization', he wrote, 'requires that the study and management of general interests should be left in the hands of an authority which is sufficiently enlightened by its positive knowledge to apply it, sufficiently elevated to take a general view and be impartial, and sufficiently strong to triumph over petty resistances'.[21]

This authority was to be acquired through the primacy of *experimentation*. Indeed, one of the key aims of the experimental method was to replace personal authority with objective, scientific authority. The physiologist Claude Bernard, for example, argued that the experimental method was essential to realize the liberal values of freedom of mind and thought. It not only had to reject the authority of philosophy and theology, but also personal scientific authority. 'The experimenter', he wrote, 'expresses his humility by renouncing the claim to personal authority; for he doubts even his own knowledge and subjects human authority to that of observation and the laws of nature'.[22] The spirit of experimentation, argued Bernard, is opposed to any uncritical acceptance of the authority of academic or scholastic sources. *Observation* – that is, direct experience – was the only

---

*Founders of the Republic: Five Studies in Nineteenth-Century French Political Thought* (Oxford, 2001), Andrew Wernick, *Auguste Comte and the Religion of Humanity: The Post-Theistic Program of French Social Theory* (Cambridge, 2001).

19   Auguste Comte, *A General View of Positivism*, 2nd ed. (1908), p. 35.

20   Ibid., 48.

21   Emile Littre, 'La Centralisation', Journal des Débats, 7 and 11 October 1862. Cited in Hazareesingh, *Intellectual Founders of the Republic: Five Studies in Nineteenth-Century French Political Thought*, p. 68.

22   Claude Bernard, *Introduction À L'étude De La Médecine Expérimentale* (1865), cited in Patrice Debré, *Louis Pasteur* (Baltimore, 1998), p. 346.

legitimate authority in the experimental method. Indeed, the experimental method subscribed to epistemological ethics of objectivity: in other words, an attempt to remove human subjectivity altogether from the process of knowledge production.[23]

Positivism was a key participant in the broad biopolitical social trend where 'the traditional moral categories that had dominated discourse about unconventional social behaviour were being eroded and replaced in a piecemeal fashion by more technical ones based on medicine and biology'.[24] Medical science had been acquiring increased authority since the eighteenth century, when 'the doctor [became] the great advisor and expert, if not in the art of governing, at least in that of observing, correcting, and improving the social "body" and maintaining it in a permanent state of health'.[25] After major health disasters such as the cholera outbreaks of 1832 and 1849, medicine and public hygiene had become ever more central to the art of government, inspiring 'the invention and the basic administrative knowledge and techniques of modernity'.[26] It was with the Pasteurian revolution, however, that the emerging authority of scientific experimentation for conducting social life was conclusively consolidated.

Pasteur's proof against the 'spontaneous generation' theory of the origins of life – demonstrating that life does not spontaneously emerge in a nutrient rich but sterile environment – was a blow to religious authority that was almost as powerful as the Darwinian revolution. Christian theology had enthusiastically embraced the spontaneous generation theory of life. Pasteur's discovery of the microbe, however, conclusively demonstrated that neither life nor disease can emerge spontaneously. Disease, he showed, is often caused by bacteria – and in doing so, he transformed understandings of the social and triggered a revolution in the public hygiene movement. Bruno Latour's celebrated account of the Pasteurian revolution analyzes how a 'new and hitherto unknown scientific authority' was created and quickly gained grip over diverse institutions and discourses across French society.[27] Latour describes Pasteur's success in terms of his ability to focus the political power of the existing hygiene movement, whose boundaries had been until now extremely vague and its energies dissipated in thin networks. Pasteur's discovery of the microbe enabled precisely the kind of reorganization of society that the political project of the hygiene movement was oriented towards. It enabled a redefinition of society, recasting the social

---

23   See the marvelous history of objectivity as a distinctive form of epistemological ethic in Lorraine Daston and Peter Galison, *Objectivity* (New York, 2007).

24   Nye, *Crime, Madness & Politics in Modern France: The Medical Concept of National Decline*, p. 61.

25   Michel Foucault, 'The Politics of Health in the Eighteenth Century', in Paul Rabinow (ed.), *The Foucault Reader* (Harmondsworth, 1984), p. 284.

26   Nikolas Rose, 'Medicine, History and the Present', in Colin Jones and Roy Porter (eds), *Reassessing Foucault: Power, Medicine, and the Body* (London, 1994), p. 55.

27   Bruno Latour, *The Pasteurization of France*, trans. Alan Sheridan and John Law (Cambridge, MA, 1988), p. 57.

link as being made up everywhere of microbes. Pasteurians and hygienists could now 'redefine what society is made up of, who acts and how, and ... become the spokesmen for these new innumerable, invisible, and dangerous agents'.[28] Pasteur demonstrated the achievements of his experimental method through dramatic public experiments. As Latour puts it, 'Pasteur's genius was in what might be called the *theatre of the proof*'.[29] The problem of the hygienists was now how 'to get people's minds used to submitting to the tutelary yoke of this new authority'.[30]

Late nineteenth-century experimental authority, which supported the increasingly technocratic government-by-expertise of the late nineteenth and early twentieth centuries, was based on developing positive knowledge about the immanent forces of matter.[31] This authority derived, firstly, from their ability to make the invisible visible – to reveal a whole new society, a world of threatening living beings lying beneath the surface of everyday life. It derived, that is, from their capacity to augment experience by amplifying otherwise imperceptible biological and social forces. Secondly, this authority derived from science's ability to ground the present in a future that could now be controlled. Rather than anchor the present on past traditions, it was becoming possible to ground it on an experimental method that promised to create new and more stable links to a perfectible future. Finally, this ability was dependent on scientists' ability to *distance* themselves from the world. Pasteur's scientific method was based on the use of laboratories which, by being carefully sealed off from the outside world, could create highly artificial environments in which microbes could flourish and hence become visible. The authority of the hygiene movement, then, derived from science's mysterious power to create spaces that were completely removed from the everyday world, and to set themselves up as the only people qualified to act as intermediaries between one space and the other.

Related to but distinct from this model of experimental authority, however, another diagram of authority can be identified, one that will receive more sustained focus over the rest of this study. This was a form of *experiential* authority that looked to the dynamics of embodied experience to lend authority to new kinds of truth claim.

## Aura and Authority

Alongside these competing forms of traditional and experimental authority, it is possible to identify a third form of authority that I will refer to as *experiential*

---

28    Ibid., 39.
29    Ibid., 85.
30    Rochard, 'L'Avenir de L'Hygiène', *Revue Scientifique* 24(9) (1887).
31    On the modern rule-by-experts, see Timothy Mitchell, *Rule of Experts: Egypt, Techno-Politics, Modernity* (Berkeley, CA, 2002).

authority.[32] This form of authority, whilst being entirely compatible with the other two, could also come into tension with them. It was these points of instability, as we will see, that Montmartre artists attempted to exploit.

In order to conceptualize the authority of experience in the nineteenth-century city, Walter Benjamin's famous analysis of the modern experience economy remains an indispensable reference point. Benjamin's writings on the role of commodification in the transformation of the structure of experience, as well as on the urban experience as a series of alienating 'shock experiences', open up a powerful diagram of the economies of embodied experience in the nineteenth-century city. As has been exhaustively analyzed elsewhere, Benjamin documents a fragmentation of experience in modernity, from long, integrated experience (*Erfahrung*) into streams of endlessly repeating, momentary, isolated experiences (*Erlebnisse*).[33]

Benjamin's famous theory of the withering of 'aura' in modernity, I wish to argue, can be fruitfully read as an account of a transformation in the techniques and technologies of authority. According to Benjamin, the commodification of experience, which reduces qualitative change to quantitative change (an endless repetition of the same), results in a destruction of 'aura' – the bewitching power of religious symbols, magic taboos and, in particular, unique stories and artworks, all of which rely on the translation and qualitative transformation of shared experience over time.[34] Genuine aura has been replaced by an obfuscatory 'phantasmagoria', the bewitching atmosphere of the commodity. This has a very different spatio-temporal structure to that of genuinely auratic objects. Whereas aura is characterized in terms of 'a strange weave of space and time: the unique appearance or semblance of distance, no matter how close it may be', the spatio-temporal structure of the commodity (and, insofar as it has itself become commodified, the modern metropolis) is one of endless repetition – a perpetual

---

32 On experiential authority, see also Claire Blencowe, Julian Brigstocke and Leila Dawney, 'Authority and Experience', Special issue of *Journal of Political Power* 6(1) (2013), Leila Dawney, 'The Figure of Authority: The Affective Biopolitics of the Mother and the Dying Man', *Journal of Political Power* 6(1) (2013), Tehseen Noorani, 'Service User Involvement, Authority and the "Expert-by-Experience" in Mental Health', *Journal of Political Power* 6(1) (2013).

33 See, for example, Claire Blencowe, 'Destroying Duration: The Critical Situation of Bergsonism in Benjamin's Philosophy of Modern Experience', *Theory, Culture & Society* 25(4) (2008), Julian Brigstocke, 'Artistic Parrhesia and the Genealogy of Ethics in Foucault and Benjamin', *Theory, Culture & Society* 30(1) (2013), Susan Buck-Morss, *The Dialectics of Seeing: Walter Benjamin and the Arcades Project* (Cambridge, MA, 1989), Howard Caygill, *Walter Benjamin: The Colour of Experience* (London, 1998).

34 See Walter Benjamin, 'On Some Motifs in Baudelaire', in Hannah Arendt (ed.), *Illuminations* (London, 1999), Walter Benjamin, 'The Storyteller', in Hannah Arendt (ed.), *Illuminations* (London, 1999), Miriam Hansen, 'Benjamin's Aura', *Critical Enquiry* 34 (2008).

production of 'novelties' between which there are no qualitative differences.[35] Commodification destroys the experience of temporal and spatial distance (and hence difference).

Benjamin explicitly sets out his account of aura via a narrative of a loss of authority. In his 'Work of Art' essay, he defines the authority of traditional art as deriving from its authenticity, which is secured by its place in tradition.

> The authenticity of a thing is the quintessence of all that is transmissible in it from its origin on, ranging from its physical duration to the historical testimony relating to it. Since the historical testimony is founded on the physical duration, the former, too, is jeopardized by reproduction, in which the physical duration plays no part. And *what is really jeopardized when the historical testimony is affected is the authority of the object.* One might encompass the eliminated element within the concept of aura, and go on to say: what withers in the age of the technological reproducibility of the work of art is the latter's aura.[36]

For Benjamin, however, this is not just an argument about art. Rather, 'This process is symptomatic: its significance extends far beyond the realm of art'.[37] Indeed, if we take seriously Benjamin's comment elsewhere that 'genuine aura appears in all things, not just in certain kinds of things, as people imagine', then it can be seen that the relation he draws between the loss of aura and the crisis of authority has a much wider field of application.[38]

Benjamin's account of the Paris Arcades exposes the new kinds of authority that objects, and the spaces in which they are displayed, start to exert through commodification. Commodities, like artworks, are invested with a kind of magical power, and Benjamin describes the Arcades as 'temples of commodity capital'.[39] However, Benjamin tells us, the spells exerted by commodities are not genuinely auratic. The commodity (unlike an auratic object) does not have the ability to return the viewer's gaze. To help explain the magical authority of the commodity, Benjamin cites Marx's observation that 'value ... converts every product into a social hieroglyphic'.[40] Commodities hide the labour (and hence the social relations) that went into making them; they are distanced from their production, and so seem to exert a magical power, as if they had emerged from nowhere, like a phantom.

---

35    Walter Benjamin, 'The Work of Art in the Age of Its Technological Reproducibility', in Howard Eiland and Michael Jennings (eds), *Walter Benjamin: Selected Writings, Volume Three, 1935–1938* (Cambridge, MA, 2002).

36    Ibid., 254. Emphasis added.

37    Ibid.

38    Walter Benjamin, 'Protocols of Drug Experiments', *On Hashish* (Cambridge, MA, 2006), p. 58.

39    Walter Benjamin, *The Arcades Project*, ed. R. Tiedmann, trans. K. McLaughlin and H. Eiland (Cambridge, MA, 1999), p. A2,2.

40    Ibid., X4,3.

Value wraps commodities in veils, shielding their nature and effects from straightforward experience.[41] This is a kind of *parody* of auratic power. Whereas aura is based on an experience on distance, no matter how near the object may seem, the (seemingly) magical power of the commodity is based on an experience of presence – the appearance of a seductive newness, no matter whether it is any different to what preceded it. Like auratic objects, commodities perform a simultaneous distancing and presencing. 'The *contemplative* attitude associated with the auratic artwork is converted into the *covetous* attitude that is applied to the stock of commodities'.[42] Commodities draw consumers near, enticing us to possess them and to incorporate them into our fantasies of a better life; but they also withdraw from us, because they cannot sate our desire, and always leave us wanting more. This gives the commodity a pseudo-auratic power: a new authority with a very different spatio-temporal structure. Whereas the authority of auratic objects connects subjects to a collective history, however, the authority of the commodity imprisons dwellers within a static temporality: an endless succession of intense but fleeting presents.

> In view of the fact that the objects of this world have lost their constant meaning fixed by tradition, 'authentic' experience become[s] privatized, transformed into an incommunicable inward event. With the disintegration of the traditional organization of experience, of the social cadres of memory, it acquires a shock-like instantaneity.[43]

Drawing on Benjamin's critical analysis of the relationship between authority, experience and the commodity, we can start to understand how the colonization of modern urban space by the commodity form leads to the proliferation of new forms of privatized and individualized 'experiential authority', where experience becomes the most trust guide for action. Behaviour becomes indexed, through capital, to attempts to accumulate more intense, fleeting experiences. In fact, further support for this approach can be drawn from one of the most celebrated French intellectuals of the 1880s: the sociologist Jean-Marie Guyau. Guyau elaborated a theory of experience which located the source of *all* legitimate authority in the dynamics of embodied experience. Rather than connecting this (as Benjamin would later do) to the dynamics of capital in modernity, however, Guyau tied it to a biopolitical logic that valued lived experience as the strongest expression of biological vitality, and hence, the best authority to guide individual behaviour and societal progress. In Guyau's writings on social authority, I wish to suggest, the

---

41   Michael Jennings, 'On the Banks of a New Lethe: Commodification and Experience in Benjamin's Baudelaire Book', in Kevin McLaughlin and Philip Rosen (eds), *Benjamin Now: Critical Encounters with the Arcades Project* (Durham, NC, 2003), p. 99.

42   Benjamin, *The Arcades Project*, cited in Gyorgy Markus, 'Walter Benjamin Or: The Commodity as Phantasmagoria', *New German Critique* 83 (2001), p. 21.

43   Ibid., 18.

biopolitical roots of experiential authority in the modern city achieved its clearest theoretical expression.

## Jean-Marie Guyau and the Authority of Experience

Jean-Marie Guyau was one of the most celebrated thinkers of 1880s France, and only his untimely death at the age of 33 stopped him becoming an even more major figure in French intellectual history. His ideas would go on to enjoy a formative influence on thinkers as diverse as Durkheim, Nietzsche and Kropotkin. As we will explore in Chapter 8, Guyau had a particularly strong influence on certain anarchist intellectuals during the period of anarchist violence known as 'propaganda by the deed' in the 1890s. It is important to introduce Guyau's ideas at this point, however, since they offer an exceptionally clear account of emerging discourses around experiential authority during the period under study. One important aspect of the authority of experience brought up by Guyau is that the roots of this authority are not just economic and technological, but also biopolitical. Experiential authority, Guyau tells us, is a form of *vital* authority, an authority that draws its source from the intensities and energies of biological life.

Guyau was strongly influenced by positivism's attempt to ground authority in science, as well as Darwinian theories of evolution. However, rather than looking for a social morality that referred to external powers, Guyau imagined a new form of social authority that would be based wholly on *experiential life*. Guyau's work developed Alfred Fouillée's argument in *Critique des Systèmes de Morale Contemporains* that moral obligation derives, not from any transcendent source, but from experience itself.[44] Guyau's innovation was to link this moral experience directly to biological life. A genuinely social morality would only evolve, he argued, through an intensification of life; for the sympathy of feeling that is required for a socially harmonious society 'is explained to a great extent by the fecundity of life, the expansion of which is almost in direct ratio to its intensity'.[45] Thus, he argued, 'It is from life that we will demand the principle of morality'.[46]

Guyau, a characteristic example of the fiercely secular spirit of the Third Republic, was concerned with addressing the vacuum of social authority that many of his contemporaries feared would be left by the decline of religious authority. Religion, he insisted, is 'simply a mythical and sociomorphic theory of the physical universe'; it is a way of explaining the world that has been superseded

---

44   Alfred Fouillée, *Critique des Systèmes de Morale Contemporains* (Paris, 1883), Robert Good, 'The Philosophy and Social Thought of Alfred Fouillée' (PhD thesis, McGill University, 1993).

45   Jean-Marie Guyau, *A Sketch of Morality Independent of Obligation or Sanction*, trans. G. Kapteyn (London, 1898), p. 70.

46   Ibid.

by modern science.[47] 'What primitive peoples were in search of when they built up their different religions was an explanation, and the least surprising explanation possible, the explanation most in harmony with their rude intelligence'.[48] Far from leaving a moral vacuum, Guyau insisted, '[e]nfeeblement of religious instinct will set free, for employment in social progress, an immense amount of force hitherto set aside for the service of mysticism'.[49]

In order to develop this argument, Guyau set out a theory of the subject as a hybrid assemblage of conflicting powers. For Guyau, every self is a composite of multiple powers: 'Our ego is but an approximation, a kind of permanent suggestion. It does not exist, it is in the process of making, it will never be complete'.[50] These powers have developed slowly over time as habit, whether those of the individual or the 'race'.[51] Habits, and hence powers, arise after organisms have successfully adapted to a certain environment. Thus to speak of 'power' is to speak of an organism's 'pre-established, constitutional adaptation, an aptitude ready to be awakened and translated into actions'. This ontology of the subject enabled Guyau to argue that every voluntary act is the product of a kind of internal, unconscious struggle for life.

> There is no completely voluntary act – or what comes to the same thing, no completely conscious act – which is not accompanied by the sense of victory of certain internal tendencies over others, and, consequently, of a possible struggle between these tendencies, and therefore of a possible struggle against them.[52]

The will, therefore, is a form of vital power. Indeed, it is an expression of the strongest, most vigorous, power. The will is simply the greatest intensity of organic power.

This theory of will as power enabled Guyau to argue that obligation or duty should not be conceived as a constraint on action, in the manner of religious morality. Rather, the feeling of obligation is a *positive experience* of a compulsion to act. Obligation is a form of intensely lived 'volition', a volition which, far from being freely chosen, is the product of an intense conflict of internal, unconscious, biological powers.

This enabled Guyau to make direct lived experience the principal source of his vision of a radical new form of social authority. Authority, for Guyau, was to derive from an amplification or intensification of *inner* experience (rather

---

47  Jean-Marie Guyau, *The Non-Religion of the Future: A Sociological Study* (New York, 1962), p. 78.

48  Ibid.

49  Ibid., 235.

50  Jean-Marie Guyau, *Education and Heredity: A Study in Sociology*, trans. W. J. Greenstreet (1891), p. 65.

51  Ibid., 47.

52  Ibid., 61.

than from an augmentation of experience through tradition, as with transcendent models of authority). For this reason, he placed high value on categories such as risk and peril, where life is raised to its greatest intensity. Those who confront risk and peril acquire authority because they have moved outside their safe milieu and hence force themselves to adapt and grow stronger. Guyau insisted that far from leaving society groundless, the decline of religious authority makes room for life immanently to create its own ground. 'Life', he wrote, 'is a kind of *gravitation upon itself*'.[53] Far from engendering a society of selfishness and greed, submitting to the authority of inner experience would enable an outpouring of social sympathy capable of binding society together in non-coercive ways. Since the essence of life is to reproduce itself, he believed, this means that the intensification and amplification of life is inevitably accompanied by a vast expenditure of charity and social sympathy. Eventually, individual experience will achieve perfect harmony with social solidarity, meaning that obedience to the authority of lived experience will result in the most generous, life-affirming and sympathetic forms of human sociability.

Guyau captured here something of the architecture of an emerging form of authority where obedience would not be tied to stable and permanent foundations and traditions, but to the inherently unstable and dynamic nature of socio-biological life. To summarize, we might say that in Guyau's thought the only legitimate forms of authority derived, first, from forces that successfully amplified lived experience, intensifying it, making it more powerful, and thereby increasing the compulsion to act. Second, legitimate authority derived from life as a force that was essentially self-grounding, a force of gravity that came, not from a distant force, but from itself. Finally, legitimate authority derived from life's capacity to transcend itself, transgressing its own limits by confronting peril and in doing so learning to adapt to new environments. Because vigorous forms of life have an inexhaustible power to grow and adapt, they always remain distant from themselves, mysterious, intoxicating, and authoritative.

Although Guyau's rejection of obligation and sanction was enthusiastically embraced by anarchists such as Peter Kropotkin, his theory sat easily within the bourgeois republican values of profit, biological vitality, and imperial expansion.[54] He insisted that his morality offered a much better foundation for the protection of private property than Christianity, which was (he argued) infested with socialist ideas.[55] He saw laziness to be the worst sin of a moral system where will is power, and claimed that the most backward forms of life – savages and criminals – are also the most workshy.[56] His theory of a morality without

---

53   Guyau, *A Sketch of Morality Independent of Obligation or Sanction*, p. 81, emphasis added.

54   Peter Kropotkin, *Ethics: Origin and Development*, trans. Louis S. Friedland and Joseph R. Piroshnikoff (New York, 1924).

55   Guyau, *The Non-Religion of the Future: A Sociological Study*, p. 241.

56   Guyau, *A Sketch of Morality Independent of Obligation or Sanction*, p. 85.

obligation or sanction makes visible aspects of the *biopolitical* architecture of experiential authority – the ways in which the authority of life could be deployed to support highly normalizing, violent and oppressive social structures. Guyau's theory exemplified the ways in which biopower travelled, not just through technical interventions in the urban environment, but at the registers of embodied experience. Experiential authority was a variety of biopolitical authority: a form of authority that stems

> from having experienced, touched upon the limits of, life. Biopolitical authority is not, then, exactly the rule of scientific truth – or of a despotic, totalising life force, collectivisation or sovereign. It is, rather, the power and attraction that congeals around a diversity of performances and manifestations of experiencing life. To be biopolitically authoritative is to mediate experience of life, to be a conduit to the force by which life (objectivity) pushes back. To know life, to make life manifest, to make a promise that life is real ... to provide a link to life is to generate biopolitical authority.[57]

Yet the link between authority and corporeal experience, as we will see in later chapters, was complex, ambiguous and heterogeneous. Experiential authority permitted new forms of *resistance* as well as supporting the continuing bourgeois conquest of the city. In the final section of this chapter, I wish to unpack some elements of this counter-hegemonic experiential authority, and the agenda that this set for the bio-cultural politics of following decades, through a reading of Baudelaire's poetic explorations of modern urban experience.

## The Authority of Suffering

In *The Use of Pleasure*, Michel Foucault suggests that Benjamin's writings on Baudelaire could usefully be re-read as a contribution to a genealogy of 'arts of living': that is, to a history of the techniques for stylizing novel relationships with moral and scientific authority and expertise.[58] Drawing on Kant's characterization of Enlightenment as the rejection of 'alien guidance' and external authority, Foucault suggests that Baudelaire articulated an exemplary ethical approach to modern authority, not simply celebrating the present moment but attempting to imagine it otherwise: that is, at once paying close attention

---

57  Claire Blencowe, 'Biopolitical Authority, Objectivity and the Groundwork of Modern Citizenship', *Journal of Political Power* 6(1) (2013), p. 20. See also the Conclusion of Claire Blencowe, *Biopolitical Experience: Foucault, Power and Positive Critique* (Basingstoke, 2012).

58  Michel Foucault, *The Use of Pleasure: The History of Sexuality, Volume Two*, trans. R. Hurley (London, 1992), p. 11. For a detailed discussion of this, see Brigstocke, 'Artistic Parrhesia and the Genealogy of Ethics in Foucault and Benjamin'.

to what is real and also trying to violate that reality.[59] If we are to follow Foucault's reading, then, Baudelaire's writings can usefully be interpreted in terms of their exploration of the limits and possibilities of an 'art of living' in the modern city and role of affective experience in such an art of life. What I will show in this section is that Baudelaire's life and poetry set a cultural agenda for mobilizing experiential authority in order to make visible new forms of *truth* concerning the life, violence and destructiveness of the modern biopolitical city.

We start with Baudelaire's poem 'Le Poison', which makes palpable a distinctive relationship between desire, the senses, and the flight towards death:

> Wine can invest the most disgusting hole
> With wonders to our eyes,
> And make the fables porticoes arise
> In its red vapour's gold
> That show in sunsets seen through hazy skies.
>
> Opium will expand beyond all measures,
> Stretch out the limitless,
> Will deepen time, make rapture bottomless,
> With dismal pleasures
> Surfeit the soul to the point of helplessness.
>
> But that is nothing to the poison flow
> Out of your eyes, those round
> Green lakes in which my soul turns upside-down …
> To these my dreams all go
> At these most bitter gulfs to drink or drown
>
> But all that is not worth the prodigy
> Of your saliva, girl,
> That bites my soul, and dizzies it, and swirls
> It down remorselessly,
> Rolling it, fainting, to the underworld![60]

This poem draws its energies from the shocking links it draws between desire and death: a radical counterpoint to the Hegelian equation of desire and life. According to Foucault, modern biopolitical authority is closely tied to knowing

---

59    Michel Foucault, 'What Is Enlightenment?', in Paul Rabinow (ed.), *Ethics, Subjectivity and Truth: The Essential Works of Foucault 1954–1984, Volume One* (London, 2000), p. 311, Kant, 'What Is Enlightenment?'. On 'immanent authority' in Foucault and Kant, see Julian Brigstocke, 'Immanent Authority and the Performance of Community in Late Nineteenth Century Montmartre', *Journal of Political Power* 6(1) (2013).

60    'Poison', in Charles Baudelaire, *The Flowers of Evil*, trans. James McGowan (Oxford, 1993).

and liberating *desire*.[61] Consequently, he tells us, it is desire that has been the most common object of ethical systems in modernity. In this respect, Baudelaire is characteristically modern. Baudelaire's poetry, however, rather than repeating the modern imperative to liberate desire as a form of creative life, is instead filled with images that depict desire as a dangerous narcotic, a poison whose false raptures can lead only towards death. In 'Poison', Baudelaire explicitly frames desire as a form of death; a lover's kiss sends the narrator's soul remorseless 'to the underworld'. Indeed, for Baudelaire modern woman becomes desirable only when experienced as an object, something lifeless.[62] Baudelaire approaches desire as a lifeless void. His approach to modern living focuses upon desire as the element of the self that has been most completely commodified and objectified, lost even to itself.

What was distinctive about Baudelaire's ethos of modernity, in Benjamin's reading, was his insistence on *encouraging* death rather than combatting it. He highlights Baudelaire's rage against the world and, in particular, against nature and biological life. 'Baudelaire', he tells us, 'found nothing to like about the age he lived in, and ... was unable to deceive himself about it'.[63] Baudelaire's 'violence, ... impatience, and ... anger' lies behind even his most lyrical poetry. Indeed, his attitude towards desire is frequently one of rage and violence. Perhaps the most dramatic example of this violence is his scandalous poem 'To One Who is Too Cheerful':

> ... Madwoman, I am maddened too,
> And hate you even as I love!
>
> Sometimes within a park, at rest,
> Where I have dragged my apathy,
> I have felt like an irony
> The sunshine lacerate my breast.
>
> And then the spring's luxuriance
> Humiliated so my heart
> That I have pulled a flower apart
> To punish Nature's insolence.
>
> So I would wish, when you're asleep,
> The time for sensuality,
> Towards your body's treasury
> Silently, stealthily to creep,

61   Michel Foucault, *The Will to Knowledge: The History of Sexuality, Volume One*, 2nd ed. (London, 1998).

62   Christine Buci-Glucksmann, 'Catastrophic Utopia: The Feminine as Allegory of the Modern', *Representations* 14 (1986), Susan Buck-Morss, 'The Flâneur, the Sandwichman and the Whore: The Politics of Loitering', *New German Critique* 39 (1986).

63   Walter Benjamin, 'The Paris of the Second Empire in Baudelaire', in Howard Eiland and Michael Jennings (eds), *Walter Benjamin: Selected Writings, Volume Four, 1938–1940* (Cambridge, MA, 2006), p. 60.

> To bruise your ever-tender breast,
> And carve in your astonished side
> An injury both deep and wide,
> To chastise your too-joyous flesh.
>
> And sweetness that would dizzy me!
> In these two lips so red and new
> My sister, I have made for you,
> To slip my venom, lovingly![64]

This poem dreams of a sexual violence whose more general target is natural life itself. It expresses a revolt against the natural world, as idealized in the stereotypical figure of the eroticized female body. It captures a form of desire that can only realize itself in the destruction of the living (like the desire aroused by the commodity form). Indeed, this relentless hostility to life applies equally to Baudelaire's own life. Benjamin emphasizes that rather than seeking refuge from the alienating effects of modernity, Baudelaire engages in a form of flânerie that involves exposing himself to the worst effects of urban life, so as to become capable of converting them into poetic *truths*. Becoming capable of capturing the experience of modernity requires exposing himself to its most violent extremities. Unlike the mere flâneur, Baudelaire does not just wander the city so as to harvest fleeting experiences, but engages in a kind of spatial *combat*:[65]

---

64   'To One Who is Too Cheerful'. Baudelaire, *The Flowers of Evil*, p. 89.
'Folle dont je suis affolé,
Je te hais autant que je t'aime!
Quelquefois dans un beau jardin
Où je traînais mon atonie,
J'ai senti, comme une ironie,
Le soleil déchirer mon sein,
Et le printemps et la verdure
Ont tant humilié mon coeur,
Que j'ai puni sur une fleur
L'insolence de la Nature.
Ainsi je voudrais, une nuit,
Quand l'heure des voluptés sonne,
Vers les trésors de ta personne,
Comme un lâche, ramper sans bruit,
Pour châtier ta chair joyeuse,
Pour meurtrir ton sein pardonné,
Et faire à ton flanc étonné
Une blessure large et creuse,
Et, vertigineuse douceur!
À travers ces lèvres nouvelles,
Plus éclatantes et plus belles,
T'infuser mon venin, ma soeur!

65   For an interesting parallel on Foucault's account of literature as spatial combat, see Peter Johnson, 'Foucault's Spatial Combat', *Environment and Planning D: Society & Space* 26(4) (2008).

Through all the district's length, where from the shacks
Hang shutters for concealing secret acts,
When shafts of sunlight strike with doubled heat
On towns and fields, on rooftops, on the wheat,
I practise my quaint swordsmanship alone,
Stumbling on words as over paving stones,
Sniffing in corners all the risks of rhyme,
To find a verse I'd dreamt of a long time.[66]

Baudelaire engages in a kind of urban joust, allowing the city to imprint itself on him, but then setting out to expose its false magic in poetic images. Thus Benjamin argues that 'the index of heroism in Baudelaire is to live at the heart of irreality (of appearance)'.[67] Baudelaire's exploration of the city should be viewed less as a hedonistic exploration of the city's fleeting pleasures, than as a highly *ascetic* ethical practice, one that involves exposing himself to the most destructive effects of the urban experience, so as to become worthy of capturing the true physiognomy of modernity in poetic images.

This is the significance of Benjamin's dazzling reading of Baudelaire as an allegorist. Benjamin argues that Baudelaire's exposure to the most alienating effects of the modern city make it possible for him to develop a way of writing, through allegory, *as a commodity*. Allegory is a form of expression that sucks the life out of its object and reveals it in a spatial image of decay and loss. It drains the life from its object, and in this way mimics the power of the commodity. Life, desire and subjectivity itself are reduced to skeletal form:

Her eyes, made of the void, are deep and black;
Her skull, coiffured in flowers down her neck,
Sways slackly on the column of her back,
O charm of nothingness so madly decked!
You will be called by some, 'caricature',
Who do not know, lovers obsessed with flesh,
The grandeur of the human armature.
You please me, skeleton, above the rest!

---

66   'The Sun'. Baudelaire, *The Flowers of Evil*, p. 169.
     'Le long du vieux faubourg, où pendent aux masures
     Les persiennes, abri des secrètes luxures,
     Quand le soleil cruel frappe à traits redoublés
     Sur la ville et les champs, sur les toits et les blés,
     Je vais m'exercer seul à ma fantasque escrime,
     Flairant dans tous les coins les hasards de la rime,
     Trébuchant sur les mots comme sur les pavés
     Heurtant parfois des vers depuis longtemps rêvés'.
67   Walter Benjamin, 'Central Park', *New German Critique* 34 (1985), p. 43.

Do you display your grimace to upset
Our festival of life?'[68]

Through allegorical images such as these, arresting the festival of life, Baudelaire stages an encounter with the *truth* of modern experience. In Benjamin's formulation, allegorical images reveal, not *Erkenntnis* or factual knowledge, but *Wahreit*, truth.[69] Whereas knowledge can be possessed and presented, truth is un-representable: it is a permanent exchange of extremes.[70] Thus Baudelaire's ethos of modernity commits itself to inhabiting a new intersection between experience and truth. Baudelaire's art of living requires him to imprint the truth on his own life so as to become worthy of presenting it with the greatest possible force in poetic images. In this way, Baudelaire can lend truth the affective density of a lived experience of suffering. Baudelaire's life, stamped with the experience of suffering, becomes the authority by which his poetic images gain access, not merely to representational knowledge, but to an un-representable truth that encompasses the opposite extremities of modern living.

Baudelaire's poetry mobilizes the authority of experience in a way that poses a radical challenge to the dominant consumerist ethics of liberating desire and harvesting momentary experiences. Baudelaire, by exposing himself fully to the violence and alienating shock-effects of the modern metropolis, uses the authority of experience to stage an encounter with the truth of modernity's violence, destructiveness, and corruption of experience. Experience, in Baudelaire, is the measure of an age in which the possibility for genuine experience has been all but destroyed. Experience (of suffering and rage) is the currency through which Baudelaire purchases the authority to voice polemical truths against modernity's destruction of authentic experience, memory and nature. As we will see in later chapters, this subversive mobilization of experiential authority was a formative influence on much of the radical urban culture of the following decades.

---

68   'Danse Macabre'. Baudelaire, *The Flowers of Evil*, pp. 197–200.
'Ses yeux profonds sont faits de vide et de ténèbres,
Et son crâne, de fleurs artistement coiffé,
Oscille mollement sur ses frêles vertèbres.
Ô charme d'un néant follement attifé.
Aucuns t'appelleront une caricature,
Qui ne comprennent pas, amants ivres de chair,
L'élégance sans nom de l'humaine armature.
Tu réponds, grand squelette, à mon goût le plus cher!
Viens-tu troubler, avec ta puissante grimace,
La fête de la Vie?'
69   Walter Benjamin, *The Origin of German Tragic Drama*, trans. J. Osborne (London, 1998), see Bainard Cowan, 'Walter Benjamin's Theory of Allegory', *New German Critique* (1983).
70   Willem van Reigen, 'Breathing the Aura – the Holy, the Sober Breath', *Theory, Culture & Society* 18(6) (2001).

## Conclusion

In this chapter I have explored the ways in which, during the 1870s and 1880s, the political culture of the Third Republic of France became dominated by narratives of a threefold crisis: a crisis of authority, an urban crisis, and a crisis of temporality. At stake in these debates was the role of tradition, religion and evolution in the preservation of social order and private property. Responses to these crises, I have argued, can be separated into three 'diagrams' of social and cultural authority. First, 'transcendent' authority was a form of authority favoured by the royalists and social conservatives, and referred to religion, tradition, and ideals of rural community. Its paradigmatic figures were the father, the notable, and the priest. It derived authority from an augmentation of experience through the accumulation over time of wisdom, knowledge and wealth. It was based on the certainty of truths that were revealed by God through his appointed intermediaries.

Second, 'experimental' authority aimed to locate authority, not in notables and priests, but in the impersonal expertise of scientists. The experimental method was intended to replace personal authority with wholly objective, scientific authority. Doing so, however, required *dramatizing* scientific experiments, creating a 'theatre of truth' in which the new authority of science could be demonstrated through spectacular displays of scientific achievement.

Finally, 'experiential' authority aimed to create order by indexing social life to the authority of corporeal experience. The increasing emphasis over the course of the nineteenth century on inhabiting urban space as a storehouse of momentary, intense experiences led to an increasing valorization of the power of experience to help order society. Theorists such as Guyau looked to biological life to find the principle of morality and social authority, and saw corporeal experience as the most intense expression of biological life. Experience, as the most authentic expression of life and vitality, was to become the most powerful form of biopolitical authority in the modern metropolis.

Finally, however, 'experiential authority' could also be mobilized to *challenge* dominant formations of biopolitical authority in the late nineteenth-century city.[71] In the writings of Baudelaire, for example, we see an attempt to ground the authority of violent challenges to the nature of modern life in the alienated, fragmented, irredeemably damaged nature of modern experience. In the following chapter of this study, we will see how such a project of contesting biopolitical authority through the invention of new, rebellious forms of bio-cultural authority were enthusiastically adopted by the bohemian artists and political activists who took up residence in the working-class area of Montmartre during the last 20 years of the nineteenth century.

---

71   On experiential authority and political contestation, see Naomi Millner, 'Routing the Camp: Experiential Authority in a Politics of Irregular Migration', *Journal of Political Power*, 6(1), 2013.

# Chapter 4
# Upsetting the Festival of Life: Humour and the Politics of Ressentiment

## Introduction

In the previous chapter, I described a threefold model of discourses of authority in late nineteenth-century France, observing the increased salience of 'experiential' authority in social, cultural and political life. Bodily, lived experience was becoming a viable competitor to the authority of the church and the state as a guide for action. Experience became seen by some as the source of a new form of distinctively modern authority that would protect the hegemony of the bourgeois, imperialist state; others, however, tried to mobilize the authority of experience in order to voice dissent against modernity's corrosive effects on nature and biological life. Following the example of Baudelaire, it was the cultural radicals taking up residence in Montmartre during the 1880s and 1890s who explored this aspect of experiential authority with the greatest intensity. In order to do so, they used humour as a way of ridiculing and challenging dominant authority structures and exploring new forms of experiential life.

This chapter explores the emergence of a novel relationship between authority, humour and biological life that started to be practiced in late nineteenth-century Montmartre. Following the Paris Commune of 1871, Montmartre became a powerful spatial figure of urban autonomy, cultural critique, and bohemian anti-authoritarianism. Amongst the opponents of the Commune, the name 'Montmartre' also crystallized fears concerning social degeneration and diseased, uncontrolled biological life. The following section explores how one early Montmartre journal, *L'Anti-Concierge*, used humour to voice discontent about the ongoing housing crisis, which had been one important source of the popular discontent that released itself explosively in the 1871 civil war. Noting the violence and rage that is easily discernible in this publication, the chapter moves on to consider the role of humour as a cultural vehicle for expressing forms of negative affect or 'ressentiment'. Yet the slightly later humour of Montmartre's cabarets, however, I go on to argue, considerably complicated this picture by taking natural vitality as the *object* of its humour. The vitality of humour became both the medium and the message. Montmartre artists and writers started to invent performance styles that parodied positivist dramaturgies of truth. With the collaboration of high level politicians as well as police agents, Montmartre emerged as a new site of experimental freedom within the Third Republic.

**Figure 4.1**    **Poster for the 'Vache Enragée' procession, 1897.**
             **Bibliothèque Nationale de France**

**Symbol of Revolt**

One event stands out as the pre-eminent symbol of the fin-de-siècle crisis of authority: the Paris Commune of 1871. Although Marxists and the anarchists of the 'Anti-Authoritarian International' disagreed about the nature of and need for authority in revolutionary uprising, both at least shared a wholesale rejection of the established authority of God and State. The Commune was seen as a powerful challenge to established authority and a utopian experiment with building new forms of municipal autonomy.

On 26 March 1871, a week after the start of the Paris Commune uprising, the first edition of a periodical called *Le Mont-Aventin: Echo des Buttes Montmartre* went to press. Only two editions of the journal were published, and it is now largely forgotten. Its significance lies in its place as the first of a flood of small, independent journals that were published in Montmartre after 1871 – over 50 periodicals were established in the area between 1871 and 1891.[1] *Le Mont-Aventin* adopted a rhetorical appeal to authority that was repeated, amplified and modified countless times by subsequent Montmartre publications. It proclaimed Montmartre as the symbolic home of the Commune. Combining genuine revolutionary fervour with a hyperbolic celebration of the role of Montmartre in the uprising, it at once celebrated and parodied the language of the 1789 revolution.[2] As an article on the opening page, titled 'La Colère du Mont-Aventin', proclaimed: 'It is from Mont-Aventin [Montmartre] that the signal for revolution was launched; it is from the heights of this Olympus that workers' anger rolled down with a din into all the neighbourhoods of Paris, enflaming the hearts of every citizen'.[3]

The comparison between Mount Olympus and the small hill of Montmartre reactivated but playfully mocked the French revolutionary tradition's habit of drawing on the authority of ancient Rome. Yet the demands it voiced were serious enough:

> The people are fed up of twenty years of monarchy, twenty years of lies, thefts, scandals, oppression and robberies. They want to re-conquer their rights; they

---

1    Maurice Artus, 'Essai de Bibliographie de la Presse Montmartoise: Journaux et Canards', *Bulletin de la societé d'histoire et d'archaeologie des IXè et XVIIIè Arrondissements: Le Vieux Montmartre* 49–50 (1906).

2    The name 'Mont-Aventin' was one of Montmartre's nicknames, named after one of the Seven Hills of Rome that lay just outside the city's sacred boundary. The hill was first known as a volatile working class district of Rome, but later became an aristocratic area. The French revolutionary tradition had long modelled itself on Ancient Rome. As Walter Benjamin puts it: 'The French Revolution viewed itself as Rome reincarnate. It cited ancient Rome exactly the way fashion cites a bygone mode of dress'. Walter Benjamin, 'Theses on the Philosophy of History', in Hannah Arendt (ed.), *Illuminations* (London, 1999), p. 261.

3    'C'est du Mont-Aventin qu'est parti le signal de la Révolution, c'est du haut de cet Olympe que, roulant avec fracas, la colère populaire s'en est allée dans tous les quartiers de Paris, enflammant tous les cœurs de citoyens'. Anon., 'La Colère Du Mont Aventin', *Le Mont-Aventin: Organe Quotidien de la Fédération Républicaine* 1871.

want municipal franchise, free elections and free revocation of their leaders".
When a people rise up with a sacred goal, when a people, having for twenty years groaned under the cane of a despot, rise up and summon the strength to claim their unprescribable rights, inalienable and sovereign, can one in good faith accuse them of wanting to foment civil war, when all hearts blessed with true citizenship are ready to unite?[4]

The journal painted a picture of a violent outpouring of revolutionary energy, down from the godly heights of Montmartre-Olympus to the mortal city below it. This proved a highly durable image. It was repeated endlessly within Montmartre over the following decades (see for example Figure 4.1). But some aspects of it were also adopted by the anti-Communard writings which caricatured the Commune as a grotesque outpouring of uncontained, corrupt and uncivilized natural forces. The energy of the revolution was represented as a violent flow of diseased and decayed life. In popular representations, Communards were feminized and bestialized, in particular through the very widespread images of Communard 'pétroleuses': gangs of women incendiaries alleged to have deliberately set fire to the city during the suppression of the revolt. Georges Clemenceau, at the time the mayor of Montmartre, described the outbreak of the revolt by agents who were inhuman and pathologically deranged:

> All were shrieking like wild beasts without realizing what they were doing.
> I observed then that pathological phenomenon which might be called blood lust. A breath of madness seemed to have passed over this mob ... women, dishevelled and emaciated, flung their arms about while uttering raucous cries, having apparently taken leave of their senses.[5]

Similarly, the poet Théophile Gautier described how: 'From opened cages rushed ... the gorillas of the Commune'.[6] The leader of the provisional government, Adolphe Thiers, referred to the revolt as 'a nervous disease' and a 'moral infection'.[7]

---

4  "'La peuple a assez de vingt ans de monarchie, de vingt ans de mensonges, de vols, de scandales, d'oppressions et de soufflets. Il veut reconquérir ses droits, il veut ses franchises municipales, la libre élection et la libre révocation de ses chefs." Quand un peuple se soulève dans un but aussi sacré, quand un peuple, après avoir gémi durant vingt années sous la férule d'un despote, se soulève et se fortifie pour la revendication de son droit imprescriptible, inaliénable et souverain, peut-on de bonne foi l'accuser de vouloir fomenter la guerre civile, quand tous les cœurs doués du véritable civisme sont prêts à s'unir?', ibid.

5  Cited in Gay Gullickson, *Unruly Women of Paris: Images of the Commune* (Ithaca, NY, 1996), p. 53.

6  Théophile Gautier, *Tableaux Du Siège, Paris, 1870–1871* (Paris, 1872), pp. 372–3.

7  Gullickson, *Unruly Women of Paris: Images of the Commune*, p. 53, Frank Jellinek, *The Paris Commune of 1871* (London, 1971), p. 99.

The revolt was also commonly read as a symptom of collective *hysteria*.[8] Emile Zola suggested 'It was … the outbreak of a morbid nervous condition, a contagious fever'.[9] It was Guy de Maupassant who explicitly named the disease that figured so prominently in anti-Communard literature: 'The Commune is nothing but Paris in the throes of a hysterical attack'.[10] This association between revolutionaries and hysteria was to become an important motif of fin-de-siècle 'decadent' counter-culture.[11]

As Colette Wilson has argued, during the years of political instability that followed the Paris Commune there seems to have been a concerted attempt to erase or negate its memory.[12] Impressionism, the dominant artistic movement of the 1870s, was largely complicit in this erasure.[13] However, the issue was reignited in 1880 when the government offered an amnesty to the Communards who had been deported to the prison colony of New Caledonia in the South Pacific. On their return, many Communards settled in Montmartre. During the 1880s, then, the memory of the Commune, and the assertion of Montmartre's role as the spatial marker for ideals of urban revolt, autonomy and freedom, started to acquire renewed prominence in the cultural imagination of the city. Montmartre stood out as a symbol of independence and claims for spatial justice. It was this symbolic legacy that the bohemian and anarchist groups who gathered in Montmartre during the 1880s and 1890s capitalized on.

## L'Anti-Concierge

Following the amnesty and repatriation of the Communards from 1880, nostalgia for the brief successes of the Commune began to assert itself with particular force. The memory of the extraordinarily brutal suppression of the revolt, however, had stilled most people's appetite for a return to revolutionary violence. Moreover, the passing of a law on the liberty of the press in July 1881 guaranteed a much broader capacity to explore *cultural* expressions of dissent and discontent. From the 1880s, a large number of small, ephemeral and often short-lived publications started to be produced in Montmartre by groups of

---

8    Janet Beizer, *Ventriloquized Bodies: Narratives of Hysteria in Nineteenth-Century France* (Ithaca, 1994).

9    Emile Zola, *La Débâcle* (Oxford, 2000).

10    'La Commune n'est pas autre chose q'une crise d'hystérie de Paris'. Guy de Maupassant, *Chroniques Inédites* (Paris, 1979), p. 2: 112.

11    Charles Bernheimer, *Decadent Subjects: The Idea of Decadence in Art, Literature, Philosophy, and Culture of the Fin-De-Siècle in Europe* (Baltimore, 2001), Noel Richard, *A L'Aube du Symbolisme: Hydropathes, Fumistes et Décadents* (Paris, 1961).

12    Colette Wilson, *Paris and the Commune, 1871–78: The Politics of Forgetting* (Manchester, 2007).

13    Albert Boime, *Art and the French Commune: Imagining Paris after War and Revolution* (Princeton, NJ, 1995).

bohemians keen to challenge established authority and moral standards. These publications were offered humorous, irreverent, and artistically experimental. Yet they voiced serious discontent concerning the nature of urban living in the Third Republic.

One Montmartre periodical called *L'Anti-Concierge* explicitly tackled an issue that had been central to the 1871 Paris Commune uprising: the housing crisis.[14] As noted in Chapter 3, the political demands of the Commune had centred on housing rights and the cancellation of rents. The revolt expressed anger at a devastating housing crisis. This crisis started to be felt again with particular force from 1880, as France slipped into economic depression, wages decreased, and rents dramatically increased. Between 1872 and 1882 rents increased by 30 to 35 per cent at the same time as salaries decreased and 10 per cent of the working population were affected by unemployment.[15] Workers were faced with unaffordable rents for squalid lodgings. At the same time, a number of anarchist groups started to organize resistance against landlords.[16] In 1881 a 'revolutionary committee for strikes against landlords' was created by anarchists who, according to police reports, urged tenants to refuse to pay their rents, and when evicted, to assemble with their furniture and children in the street and demand to be housed in unoccupied properties owned by the state. Shortly afterwards, more practical solutions were adopted, with the group organizing help for struggling tenants to surreptitiously move out of the lodgings under the cover of night, shortly before the quarter's rent was due, preferably with all their furniture in tow.

*L'Anti-Concierge* used humour to voice aspects of the rage and anger underlying the politics of this anarchist housing action group, voicing grievances against the state of the housing endured by most of the working-class inhabitants of the city. Launched by a group of Montmartre artists, it raged against the small humiliations and violations of liberty – as well as the wider economic hardships – suffered daily by tenants. Styling itself with mock seriousness as the 'organe officiel de la défense des locataires' (official organ for the defence of tenants), its primary target was the 'concierge': employees of the landlord, hired to manage and monitor the property. These authority figures, occupying a difficult mediating position between tenant and landlord, and not much wealthier than the tenants, were hugely unpopular, along with related characters such as doormen, porters and property managers. The *Anti-Concierge* directed a great force of anger against them. The opening article of the first issue, in a daringly avant-garde typographical format, announced its programme (Figure 4.2):

---

14    See also Peter Soppelsa, 'The Fragility of Modernity: Infrastructure and Everyday Life in Paris, 1870–1914' (PhD thesis, The University of Michigan, 2009).

15    Cécile Péchu, 'Entre Résistance et Contestation: La Genèse Du Squat Comme Mode d'Action', *Université de Lausanne Travaux de Science Politique* 24 (2006).

16    See ibid.

**Figure 4.2** *L'Anti-Concierge*, vol. 1, 1881

PLEASE WIPE YOUR FEET, when they are nice and dirty, on the head of your concierge; put bugs in his butter and itching powder in his bed; sow fulminating peas in his alcove, and put pepper in his wife's snuffbox; throw him scornful glances and hurl flower pots at his figure; tie twenty furious cats and a few dozen

rabid dogs to each other by the tail and release them onto the staircase in the
night; pour salt in his coffee and sprinkle his vermicelli and tapioca with sugar;
spread slanders about him and put orange peel and apple skins in his corridor.
Finally, create all the misery for him that the indignation of the oppressed
tenant suggests to you. This is the program that *L'Anti-Concierge* submits to its
innumerable readers.[17]

The theme of the journal, then, was humorous advocacy of disobedience against
the landlords and property speculators who were making enormous profits
at the expenses of their tenants. The journal offered its own version of the
'Marseillaise':

### Tenants' 'Marseillaise'

Let's go, unhappy tenants,
Payment day has arrived;
The sinister standard of the landlords
Has been raised against us. (Repeat.)
Do you hear, gloomy orchestra,
These cruel chatterboxes mooing?
They are coming with their brushes,
To demand the money for the quarter ...
To arms, townsmen! Stand upright on the landing!
Let's strike, let's strike! Until impure blood is running in the stairwell!
Would these terrible Cerberuses
Make the law in our homes?
What? Would mercenary caretakers claim from us our rents? (repeat.)
Good God! The hands of these door-shutters rummaging in our pockets!
These pull-cords would be masters to force our doors! ...
To arms, townsmen! Stand upright on the landing!
Let's strike, let's strike! Until impure blood is running in the stairwell!
Tremble, porters, and all your wives,

---

17 'ESSUYEZ VOS PIEDS, S'IL VOUS PLAIT. Quand ils sont bien sales, sur la tête
de votre concierge, mettez-lui des punaises dans son beurre et du poil à gratter dans son lit;
semez des pois fulminants dans son alcôve, et du poivre dans la tabatière de son épouse,
jetez-lui des regards méprisants et des pots de fleurs sur la figure; attachez par la queue
vingt chats furieux et quelques douzaines de chiens hydrophobes que vous lâcherez la nuit
dans les escaliers; versez du sel dans son café et saupoudrez de sucre son vermicelle et son
tapioca; répandez sur son compte des bruits calomnieux et sur son passage des écorces
d'oranges et des pelures de pommes. Enfin faites-lui toutes les misères que votre indignation
de locataire opprimé peut vous suggérer. Tel est le programme que l'Anti-Concierge soumet
à ses innombrables lecteurs'.

Spouses worthy of their husbands;
Tremble! All your infamous plans will finally receive their due. (repeat.)
Everyone is a soldier to fight you: children, women, young and old;
You will find them all in place,
Ready to fight you!
To arms, townsmen! Stand upright on the landing!
Let's strike, let's strike! Until impure blood is running in the stairwell!
. ... [etc.].[18]

Many of the articles were devoted to documenting stories of particularly unpleasant, foolish, or depraved concierges. The tone was often highly misogynist, as with

---

18    Anon., 'Marseillaise Des Locataires', *L'Anti-Concierge: Organe Officiel de la Défense des Locataires* 1? (1881), ibid.
'Allons, malheureux locataires,
   Le jour du terme est arrivé ;
   Contre nous des propriétaires
   L'étendard sinistre est levé. (Bis.)
   Entendez-vous, lugubre orchestre,
   Mugir ces cruels pipelets
   Ils viennent avec leurs balais
   Réclamer l'argent du trimestre ...
Aux armes, citadins! Debout sur le palier!
Frappons, frappons! Qu'un sang impur coule dans l'escalier!
Que veulent ces port-calottes,
Ces édentés toujours rageant?
Pour qui ces redoutables notes,
Qui nous demandent de l'argent? (Bis)
Frères, pour nous; levons nos têtes!
Apprêtons-nous à résister ;
C'est nous qu'on ose méditer
De contraindre à payer nos dettes! ...
Aux armes, citadins! etc. etc.
Qui! d'épouvantables Cerbères
Feraient la loi dans nos foyers?
Quoi? des concierges mercenaires
Nous réclameraient nos loyers? (Bis)
Grand dieu! les mains de ces clos-portes
Dans nos poches farfouilleraient!
Des tire-cordons deviendraient
Les maitres d'enfoncer nos portes! ...
Aux armes, citadins! etc.
Tremblez, portiers, et tous leurs femmes, épouses dignes des maris ;
Tremblez! tous vos projets infâmes
Vont enfin recevoir leur prix. (Bis)
Tout est soldat pour vous combattre,
Enfants, femmes, jeunes et vieux;
Vous en trouverez jusqu'aux lieux
Contre vous tout prêt à se battre! ...
Aux armes, citadins! etc.etc.'.

the description of the concierge of 9, Place Michel as 'an old woman, a vicious animal who, in spite of her great age, dreams only of love and pornography'.[19] It was populated with mock advertisements such as the personal advert placed by 'A former murderer just released from twenty years of forced labour [who] seeks a place as a concierge. Dreadful references. Repulsive appearance'.[20]

Whilst in certain respects the humour of *L'Anti-Concierge* set the tone for much of the later urban culture of fin-de-siècle Montmartre, this earlier publication lacked the affirmative ethos of carefree humour and gaiety that would later be seen in the cabarets of the 1880s.[21] The writing of the *L'Anti-Concierge* made much more explicit the forms of anger, resentment and latent violence that underpinned much Montmartre humour. Whilst the neighbourhood quickly became an emblem of (supposedly) authentic Gallic pleasure and 'esprit', an essential part of this pleasure was its capacity to sublimate the anger created by the routine everyday experiences of poverty, suffering and drudgery.

This latent violence of humour has in fact often been observed in philosophical and social scientific writings on the topic. For example, the ancient 'superiority' theory of humour draws attention to precisely this aspect of its power. Dating back to Plato and Aristotle, the most well-known advocate of the superiority theory is the seventeenth-century philosopher Thomas Hobbes, who wrote that 'the passion of laughter proceedeth from a sudden conception of some ability in himself that laugheth. Also men laugh at the infirmities of others, by comparison wherewith their own abilities are set off and illustrated'.[22] Humour, for Hobbes, is a celebration and assertion of power: the person who laughs is delighted to discover a particular strength or virtue in himself that is lacking in another. In this sense, he writes, laughter is the opposite of tears, which arise from the recognition of an inferiority or weakness in oneself.

Scholars of humour have rightly insisted that many forms of humour exist that do not fit this 'superiority' thesis. Nevertheless, the humour of *L'Anti-Concierge* seems a clear example of a variety of humour that was oriented towards a hostile assertion of power and superiority. Based on a reversal of established hierarchies,

---

19  'La concierge du 9, place Saint-Michel, est un vieille, bête vicieuse qui, malgré son grand âge, ne rêve que d'amour et de pornographie'. Anon., 'Echos De L'escalier', *L'Anti-Concierge: Organe Officiel de la Défense des Locataires* 1 (1881).

20  'Un Ancien Assassin qui vient de condamnation a vingt ans de travaux forcés, demande un place de concierge. Détestables références. Physique repoussant'.

21  See, for example: Phillip Dennis Cate and Mary Shaw (eds), *The Spirit of Montmartre: Cabarets, Humor, and the Avant-Garde, 1875–1905* (New Brunswick, NJ, 1996), Janet Whitmore, 'Absurdist Humor in Bohemia', in Gabriel Weisberg (ed.), *Montmartre and the Making of Mass Culture* (New Brunswick, NJ, 2001), Michael Wilson, 'Portrait of the Artist as a Louis XIII Chair', in Gabriel Weisberg (ed.), *Montmartre and the Making of Mass Culture* (New Brunswick, NJ, 2001).

22  Thomas Hobbes, 'Human Nature', in J. C. A. Gaskin (ed.), *Human Nature and De Corpore Politico* (Oxford, 1994), p. 54. On philosophical theories of humour, see, for example, John Morreall, *The Philosophy of Laughter and Humor* (Albany, NY, 1987).

it celebrated the moral and aesthetic superiority of tenants over those who exerted illegitimate authority over them. In this sense, it could be interpreted as a form of symbolic resistance, a means of affirming tenants' right to living quarters that were healthy, affordable and private. The humour of *L'Anti-Concierge* perhaps played a role in resisting and subverting established norms and hierarchies.[23]

However, this reading of *L'Anti-Concierge*'s vitriolic humour fails to explain how such a form of culture could plug into the wider assemblages of critical thought and affective communication without which it could have little productive power. A more convincing reading is to interpret such forms of humour as a momentary, bitter and reactionary compensation for the misery of social injustice and inequalities. Rather than making any serious cultural challenge to wider social, political and economic structures of domination, the writers of the journal simply found an easy scapegoat in the authority figure closest to them: the concierge. Such an interpretation sees humour, not as a form of effective opposition, but as provoking a form of laughter that 'copes with fear by defecting to the agencies which inspire it'.[24] For Theodor Adorno and Max Horkheimer, for example, humour is one of the most important weapons of capitalist culture industries, which function to destroy critical awareness by offering pleasures that create fleeting distractions from the miseries and boredom of everyday life in capitalist modernity. In the hands of the culture industry, they write, laughter becomes 'a medicinal bath which the entertainment industry never ceases to prescribe. It makes laughter the instrument for cheating happiness'.[25] Whereas in a fully reconciled society laughter has a positive and affirmative function, in mass society it becomes reduced to 'a sickness infecting happiness and drawing it into society's worthless totality'.[26] Unable to mount an effective political discourse, the humour of figures associated with the *Anti-Concierge* amounted to little more than a defeatist celebration of political impotence, rather than an expression of a form of dissent with the potential to productively resonate with effective forms of political resistance. In this sense, the laughter

---

23    On humour and resistance more generally, see, for example, Chad Bryant, 'The Language of Resistance? Czech Jokes and Joke-Telling under Nazi Occupation, 1943–45', *Journal of Contemporary History* 41(1) (2006), Chris Powell and George Paton (eds), *Humour in Society: Resistance and Control* (Basingstoke, 1988), S. B. Rodrigues and D. L. Collinson, '"Having Fun"? Humour as Resistance in Brazil', *Organisation Studies* 16(5) (1995), Scott Sharpe, Maria Hynes and Robert Fagan, 'Beat Me, Whip Me, Spank Me, Just Make It Right Again: Beyond the Didactic Masochism of Global Resistance', *Fibreculture* 6 (2005), ibid., Marjolein t'Hart and Dennis Bos (eds), *Humour and Social Protest* (Cambridge, 2007), Simon Weaver, 'The "Other" Laughs Back: Humour and Resistance in Anti-Racist Comedy', *Sociology* 44(1) (2010).

24    Theodor Adorno and Max Horkheimer, *Dialectic of Enlightenment*, trans. J. Cumming (London and New York, 1997), p. 152.

25    Ibid.

26    Ibid. On Adorno's positive account of laughter, see Shea Coulson, 'Funnier Than Unhappiness: Adorno and the Art of Laughter', *New German Critique* 34(1 100) (2007).

of the *Anti-Concierge* corresponds precisely to what Friedrich Nietzsche refers to as 'ressentiment':

> a desire to deaden pain by means of affects ... to deaden, by means of a more violent emotion of any kind, a tormenting, secret pain that is becoming unendurable, and to drive it out of consciousness at least for the moment: for that one requires an affect, as savage an affect as possible, and, in order to excite that, any pretext at all.[27]

It might be tempting, therefore, to reduce Montmartre's entire culture of carnivalesque humour to a defeatist submission to the fear of political revolt induced by the brutal violence of the 1871 civil war. Yet this is not the whole story. Later Montmartre urban culture, I propose, added a new ingredient to their humour that demands more sophisticated analysis. This element was the use of humour to reproblematize one of the most controversial social processes of late nineteenth-century society: the life of the city. Adequately understanding the cultural politics of fin-de-siècle Montmartre, I will argue, requires introducing a new axis of analysis, one that relates fin-de-siècle humour to the experience of biological life and emerging forms of late nineteenth-century bio-cultural politics.

In order to begin to conceptualize this, it is useful to turn to Henri Bergson's vitalist theory of comedy. According to Bergson, humour is a means of affirming the suppleness and creativity of biological life, and opposing life's petrification, mechanization or objectification. Humans become comic when they give the impression of being a mere thing rather than a living being. The comic arises from moments where people find themselves unable to adapt organically to their environment – such as when a man trips over a paving stone.[28] Here, 'the laughable element ... consists of a certain mechanical inelasticity, just where one would expect to find the wide-awake adaptability and the living pliability of a human being'.[29] Indeed, Bergson goes on, 'The attitudes, gestures and movements of the human body are laughable in exact proportion as that body reminds us of a mere machine'.[30]

This relationship between life and laughter, we should note, has a flip side. Humour is an affirmation of creativity, suppleness and novelty, but *at the same time* a powerful means of social control, a way of punishing a failure to conform to social norms. This is because it is not just physical misfortunes that betray a lack of organic suppleness; it is also social awkwardness – a failure to belong adequately to the group. 'Society', Bergson remarks, will

---

27    Friedrich Nietzsche, *On the Genealogy of Morals*, trans. W. Kaufmann and R. J. Hollingdale (New York, 1969), p. 127.

28    Henri Bergson, *Comedy* (New York, 1956).

29    Ibid., Chapter 2.

30    Ibid., Chapter 4.

be suspicious of all inelasticity of character, of mind and even of body, because it is the possible sign of ... an activity with separatist tendencies, that inclines to swerve from the common centre round which society gravitates ... By the *fear* which it inspires, [laughter] restrains eccentricity.[31]

Anticipating later sociological analyses of the role of embarrassment and ridicule in ensuring conformity to social norms, Bergson's analysis emphasizes that laughter is a powerful yet non-coercive means of sanctioning social non-conformity.[32]

In this sense, however, the Bergsonist theory of humour effectively suppresses forms of life that are unrecognized or unacknowledged by societal norms. Humour, that is, becomes a force that acts *against* life rather than for it. Its role is to ensure conformity. As Adorno puts it, 'The fact is that laughter ... has long become the weapon of convention against uncomprehended life, against the traces of something natural that has not been quite domesticated'.[33] Humour becomes a means of affirming conventional life by excluding and ridiculing those forms of potentially disruptive life and vitality (the kind of life that Adorno calls uncomprehended and undomesticated) – that are foreign to it. Humour thus functions, here, as a device for excluding diversity and difference.

What these arguments also imply, however, is that humour might have played a novel kind of epistemological role in making life available to knowledge and experience. Humour, I want to suggest, offered an alternative cultural vehicle for making the limits and boundaries of socio-biological life *visible* and *felt* at affective registers of experience. If humour is a means of affirming particular forms of life and excluding others, then it is also a means of identifying and testing the limits of life. In other words, far from necessarily being a mere distraction form the real world, humour can also function as a powerful technique for truth-telling. Humour, by testing the limits of life, can dramatize and critique established knowledge claims concerning the nature of healthy life. Indeed, the enthusiasm of the Montmartre counter-culture for developing new forms of humour, I wish to suggest, was closely linked to an aesthetico-political project of finding novel means for making visible the life of the city – and, perhaps, challenging accepted knowledges and experiences of urban vitality. Humour, I will argue, was mobilized not simply in a fearful affirmation of 'domesticated' life, but as a reflexive engagement with the parameters of life itself. Humour could lend the authority of embodied, vital, living experience to new truth-claims concerning

---

31    Ibid., Chapter 2, emphasis added.

32    For a more fully developed account of the sociology of embarrassment and ridicule, see Michael Billig, 'Humour and Embarrassment', *Theory, Culture & Society* 18(5) (2001), Michael Billig, *Laughter and Ridicule: Towards a Social Critique of Humour* (London, 2005), Erving Goffman, 'Embarrassment and Social Organization', *Interaction Ritual: Essays on Face-to-Face Behaviour* (New York, 1967).

33    Theodor Adorno, *Negative Dialectics*, trans. E. Ashton (London, 1973), p. 334, translation modified.

the true life of the city. Humour, as the most powerful expression of affective life, was the most effective authority through which to test the limits and thresholds of experiential life.

In order to starting exploring this hypothesis, I wish to start with a reading of some parodies of positivist scientific, 'experimental' forms of authority relation. As we will see, by parodying the positivist dramaturgy of truth, Montmartre bohemians experimented with using humour to mobilize an alternative form of authority – one based on lived, subjective experience rather than abstract scientific method – through which to lay claim to the possibility of 'other' forms of life beyond those imagined through hegemonic biopolitical discourses.

## Experimental Authority and the 'Cabaret Artistique'

At the same time as Pasteur was so successfully isolating and amplifying the hidden agents fighting against urban health and vitality (see Chapter 3), a new kind of cultural space was being invented on the slopes of Montmartre. The 'cabaret artistique' was a novel performance space where artists would come to perform and display their works, as well as exchange ideas, in a relaxed and convivial environment.[34] The most well-known example of the Montmartre cabaret was the 'Chat Noir'.[35] It formed a model that would later spread across France and throughout Europe, as far away as Barcelona, St Petersburg and Kraków.[36] Other similar cabarets quickly sprang up all over Montmartre.

What was distinctive about such cabarets was their fondness for sending up established aesthetic and literary values and debates, and their propagation of a spirit of anarchic humour, parody and satire often referred to as 'fumisme'.[37] The cabaret soon became a fashionable destination for writers, poets, musicians, and artists, who were both the clients and the entertainment, since they lost no opportunity to donate pictures, recite poetry, and play music. Something of the atmosphere of the Montmartre cabarets is captured by Charles Rearick:

> As a refuge from the workaday world, the cabaret imparted the illusions of theatre, but with much more spontaneity and interaction between audience and spectacle, clients and performers ... When not performing, the artists mingled

---

34   Lisa Appignanesi, *The Cabaret* (London, 1984), Elisabeth Pillet, 'Cafés-Concerts et Cabarets', *Romantisme*, 75 (1992), Harold Segel, 'Fin De Siècle Cabaret', *Performing Arts Journal* 2(1) (1977).

35   Mariel Oberthur, *Le Cabaret du Chat Noir a Montmartre* (Geneva, 2007).

36   Harold Segel, *Turn-of-the-Century Cabaret: Paris, Barcelona, Berlin, Munich, Vienna, Cracow, Moscow, St. Petersburg, Zürich* (New York, 1987).

37   Phillip Dennis Cate, 'The Spirit of Montmartre', in Phillip Dennis Cate and Mary Shaw (eds), *The Spirit of Montmartre: Cabarets, Humor and the Avant-Garde, 1875–1905* (New Brunswick, NJ, 1996), Whitmore, 'Absurdist Humor in Bohemia'.

with the customers; a kind of festive fraternity took the place of rigid roles and hierarchies entrenched in everyday society outside.[38]

By championing minor art forms (such as puppetry, song, comedy and shadow plays), adopting an aggressively anti-bourgeois stance, and experimenting with redrawing the boundaries between art and life, the cabarets formed an important part of an emerging avant-garde culture.[39] Through their enthusiastic resistance to moral conventions and establishment values – their attempts to rebuild life around entirely different values to those championed by the Republic's bourgeoisie – they made Montmartre into the site of an important set of social experiments with the nature and limits of the individual freedoms guaranteed by the new Republic.

Indeed, the highly experimental nature of the Montmartre cabarets invites comparison with the spaces of *scientific* experimentation that came to prominence during the same period. Drawing on the work of Alexander Vasudevan, we might diagram these performances as forms of 'experimental embodiment': that is, as 'critical aesthetic interventions that were … tasked with performing scientific experiments with their own alternative regimes of truth'.[40] Montmartre cabarets created novel forms of cultural experience in order to make visible the limits of life and, in particular, to contest hegemonic truths about the *pathological* nature of urban modernity. Motifs of disease, madness, and degeneration were central to the style and content of the cabaret performances. The cabarets were spaces in which pathologies could be performed, made visible, and experientially lived. Indeed, 'Modernity in the cabaret [was], to an astonishing extent, another word for pathology'.[41] They were biopolitical spaces in which controversial aspects of the life of the city were intensified, dramatized, parodied, and creatively reconfigured. As Rae Beth Gordon has explored, central to the performance style in the cabarets artistiques was the humorous imitation of pathological diseases such as hysteria and epilepsy.[42] Hysteria, a disease of excess and exaggeration

---

38   Charles Rearick, *Pleasures of the Belle Epoque: Entertainment & Festivity in Turn of the Century France* (New Haven, 1985), pp. 59–60.

39   Cate and Shaw (eds), *The Spirit of Montmartre: Cabarets, Humor, and the Avant-Garde, 1875–1905*.

40   Alexander Vasudevan, 'Symptomatic Acts, Experimental Embodiments: Theatres of Scientific Protest in Interwar Germany', *Environment and Planning A* 39(8) (2007), p. 1814.

41   Rae Beth Gordon, 'From Charcot to Charlot: Unconscious Imitation and Spectatorship in French Cabaret and Early Cinema', *Critical Inquiry* 27 (2001), p. 529.

42   See Mary Gluck, *Popular Bohemia: Modernism and Urban Culture in Nineteenth-Century Paris* (Cambridge, MA, 2005), Rae Beth Gordon, 'Le Caf Conc' et L'Hystérie', *Romantisme* 64 (1989), Gordon, 'From Charcot to Charlot: Unconscious Imitation and Spectatorship in French Cabaret and Early Cinema', Rae Beth Gordon, *Why the French Love Jerry Lewis: From Cabaret to Early Cinema* (Stanford, CA, 2001), Rae Beth Gordon, *Dances with Darwin, 1875–1910: Vernacular Modernity in France* (Farnham, 2009).

that was one of the most fashionable of fin-de-siècle psychological maladies, was believed to offer a window into the human unconscious.[43] It was a highly dramatic illness, and curious Parisians could watch spectacular performances of hysteria at Charcot's famous weekly lectures where his patients theatrically demonstrated their symptoms. Hysteria soon spread beyond Charcot's theatre, however. Imitation of the frantic, mechanical, puppet-like gestures of the hysteric became a routine part of fin-de-siècle popular performance. The most well-known male singer of the era, known as Paulus, developed a style which the critic Francisque Sarcey termed 'gambillard', in reference to his imitation of the dislocations and gesticulations of a puppet. Paulus adopted a frantic, disjointed movement that was seen as distinctively epileptic and puppet-like. 'He knocks back his head behind him while making his arms and hand play around at the same time. He leaps about, he winks mysteriously, he scans every crash of the cymbal with his head', Sarcey writes.[44] Georges Montorgeuil described Paulus as a

> hunter of regulated mechanical tics [...] a veritable electric battery ... Oh! Angles everywhere on this weak body; something automatic in the gestures [...] a mouth which is split in a tormented grin; an astonishing mask ending at two points: on top, the cranium, below, the chin: [...] planted by the devil on a bundle of nerves.[45]

Posters for Paulus' concerts reveal this distorted, maniacal, unnatural posture to be a key selling point for his act (Figure 4.3).

Such mechanical, inorganic gestures were repeated throughout the Montmartre cabarets. They were associated above all with the pantomime figures such as Pierrot with whom Montmartre bohemians strongly identified (see Chapter 5). Barbey d'Aurevilly described the style of the Montmartre singer Maurice Rollinat as that of a 'steel devil; sharpened steel that chills you whilst it cuts you ... he has the devil in his body ... a visionary of visions which he does not see. And it is in this way that we have a poet who is all the more modern'.[46]

---

43    See Beizer, *Ventriloquized Bodies: Narratives of Hysteria in Nineteenth-Century France*, Martha Noel Evans, *Fits and Starts: A Genealogy of Hysteria in Modern France* (Ithaca, NY, 1991), Rhona Justice-Malloy, 'Charcot and the Theatre of Hysteria', *The Journal of Popular Culture* 28(4) (1995).

44    'Il renverse la tête en arrière en faisant jouer en même temps ses bras et ses mains. Il gambade, il cligne de l'œil mystérieusement, il scande de sa tête chaque coupe de cymbale'. Cited in François Caradec and Alain Weill, *Le Café-Concert* (Paris, 1980).

45    'Chercheur de tics mécaniques réglés [...] une véritable pile électrique ... oh! les angles partout sur ce corps chétif, dans les gestes quelque chose d'automatique [...] une bouche qui se fend en un rictus tourmenté; un masque étonnant terminé par deux points: en haut, le crâne, en bas, le menton: [...] planté à la diable sur un paquet de nerfs'. Georges Montorgueil, cited in ibid., 138.

46    'Diable en acier; en acier aiguisé qui coupe et qui fait froid en coupant ... il a le diable au corps ... visionnaire des visions qu'il ne voit pas. Et c'est ainsi que nous

**Figure 4.3    Alcazar d'Eté. 'Tous les soirs original Paulus'. Poster, 1890. Bibliothèque Nationale de France**

Indeed, the Montmartois performer Maurice Macnab attempted to transform the written word itself into an epileptic fit. In a strange poem titled 'Ballad of Circumflex Accents', he writes:

> When I saw gliding over a word
> The biconvex shape
> Of a vulgar circumflex accent,
> This used to make me very confused
> As I was then just a kid
> …
> Sometimes they seem to be involved
> In incredible gymnastics,
> Adopting fantastic poses …
> Their form changes constantly
> …
> In old Gothic manuscripts,
> They cover like a lamp-shade

---

avons un poète moderne de plus'. Jules Amédée Barbey D'Aurevilly, 'Maurice Rollinat', *Le Constitutitionel*, 1 June 1882.

The five vowels in turn,
Which, under their vulture claws,
Make epileptic circles.[47]

With their jagged, angular, unnatural performance style, mimicking the style of a puppet or a clown, performers in fin-de-siècle cabarets purposefully adopted exactly the kind of inorganic postures that vitalist philosophers such as Bergson would later mark out as the essence of the comic. The Montmartre cabaret was emerging as a space for playing with and testing emerging knowledges of the limits of life in the modern city.

It is tempting, then, to make connections between cultural experiments with testing the limits of life in the Montmartre cabarets, and the kinds of experimental authority that were being deployed in theatres of truth associated with scientists such as Pasteur and Charcot. There was a strong emphasis in Montmartre cabarets on humorous *parody* of the positivist theatre of truth.[48] In doing so, the cabarets can be seen to have played a part in negotiating the new meanings and experiences of urban vitality and degeneracy that were emerging with such extraordinary force

---

47   Maurice Macnab, 'Ballade Des Accents Circonflexes', *Poèmes Mobiles* (1890).
'Lorsque je voyais sur un mot
Planer la forme biconvexe
D'un vulgaire accent circonflexe,
Cela me rendait très perplexe,
Etant alors jeune marmot.
Comme en de sombres paysages,
La chauve-souris des sabbats
Vole en rasant le sol très bas ...
Tantôt ils semblent occupés
A d'incroyables gymnastiques:
Prenant des poses fantastiques,
Ce sont alors de longs moustiques
Dont bras et pattes sont coupés! ...
Leur forme change à tout moment:
C'est un chapeau de commissaire,
Puis un capuchon débonnaire,
Une brosse de dromadaire,
Ou le fronton d'un monument!
Dans les vieux manuscrits gothiques,
Ils coiffent comme un abat-jour
Les cinq voyelles tour à tour
Qui, sous leurs griffes de vautour,
Font des rondes épileptique'.

48   For a critique of social scientists' fondness for over-extending the concept of the 'laboratory', see Michael Guggenheim, 'Laboratizing and De-Laboratizing the World', *History of the Human Sciences* 25(1) (2012). The actual scientific laboratory is a highly controlled, experimental environment. It is 'a purified and simplified world', a world reduced to a manageable scale. Michel Callon, Pierre Lascoumes and Yannick Barthe, *Acting in an Uncertain World: An Essay on Technical Democracy* (Cambridge, MA, 2009), p. 50.

from the new scientific laboratories of the late nineteenth century. The Montmartre cabaret, I am proposing, acted as a popular cultural vehicle for examining new and deeply disturbing knowledges of urban life and disease, by parodying and creatively reworking hegemonic spaces of scientific discovery. Indeed, the Montmartre cabaret can be interpreted as a kind of parody of the scientific laboratory itself, one that enabled the knowledges emerging from the opaque space of the laboratory to be experienced and reinterpreted.

This parody with the scientific laboratory functioned in at least three ways. First, just as experimental laboratories set out to create ideal environments for nurturing pathological elements under controlled conditions, the Montmartre cabarets represented themselves as spaces in which diseased, decadent or corrupt vital energies, rather than being repressed (as they were in the everyday spaces of normal life), could instead be accentuated, exaggerated, and hence made visible and *experienced*. Montmartre was 'an environment in which decadence could be observed, lived, condemned, or shared'.[49] In this way, by engineering intense experiences of pathological life, it became possible to experiment with new experiences of the life of the city. Parodic performances of hysteria and other perceptual diseases (see Chapter 7) created alternative theatres of truth in which dominant framings of biological health might be tested and contested.

Second, just as the authority of the scientific discoveries of figures such as Pasteur and Charcot derived from elaborate sets of performances that dramatized his new discoveries about the nature of life, the Montmartre cabarets offered a space in which emerging knowledges of life, vitality and disease could be displayed in spectacular and entertaining cultural performances. These parodic performances, however, also made it possible for audiences to *participate*, for a short while and in a safe environment, in modern experiences of diseased social and cultural life. Part of the reason for the huge popularity of the Montmartre cabarets arguably derived from their ability to invent highly participatory cultural experiments in which audiences could test emerging socio-biological discourses concerning the city at the level of their own lived experiences. In Montmartre's cabarets, audiences found a safe environment in which they could participate in the outer limits of the experiential life of the city.

Finally, just as Pasteur's laboratories carefully contained the production of pathological life within discrete and controlled spaces, Montmartre cabarets emphasized their isolation and separation from the rest of the city. We will explore in more detail the ways in which Montmartre cabarets played on this spatial separation in Chapter 5. Here, however, it is interesting to observe the extent to which the government and police, through their careful monitoring of the Montmartre cabarets, unwittingly participated in this parody of the controlled scientific experiment. By policing Montmartre as an experimental biopolitical

---

49   Richard Thompson, 'Introducing Montmartre', in Richard Thomson, Phillip Dennis Cate and Mary Weaver Chapin (eds), *Toulouse Lautrec and Montmartre* (Washington, 2005), p. 67.

space for governing through freedom, the police, as we will see in the following section, themselves contributed to a performance of the Montmartre cabaret as a space not wholly dissimilar to a laboratory in which dangerous biological forces could be allowed to multiply and make themselves visible whilst being contained within a spatially delimited, safe space.

## Contained Freedom

The police were inevitably troubled by the emergence of these spaces of disorderly, diseased, troublesome urban life. Rather than controlling them more tightly, however, there is clear evidence to suggest that police took the exact opposite approach, offering the Montmartre cabarets an unprecedented degree of freedom. In a fascinating letter from the Prefect of the Police to the Minister for Education and the Arts, the Prefect acknowledges the extraordinary liberty that had been granted to Montmartre cabarets. 'When the Prefecture of the Police wanted to prosecute', he writes,

> it found itself in the presence of such powerful influences that it had to give up submitting the Chat Noir cabaret, which was supported energetically by numerous notables in politics, letters and the arts, to the common system – Among its protectors we will quote in particular [...] the President of the Council himself who, on 20 December 1888, ordered over the telephone for M. Salis to be treated with leniency.[50]

Emphasizing that such freedom had only been granted to Montmartre cabarets, and not to the cabarets in the student-filled Latin Quarter, the Prefect goes on:

> if we could, without too many disadvantages, exercise a certain degree of tolerance towards cabarets located in a special artistic milieu like that of Montmartre, there were real dangers in exercising the same tolerance towards those frequented by young students.[51]

---

50   Letter, 11 March 1887, Archives Nationales, F. 21 338. '[Q]uand la Préfecture de Police voulut poursuivre, elle se trouva en présence d'influences tellement puissantes qu'elle dut renoncer à soumettre au régime commun le cabaret du Chat Noir, soutenu énergiquement par de nombreuses notabilités de la politique, des lettres et de l'art. – Parmi ses protecteurs nous citerons notamment [...] le Président du Conseil d'alors lui-même qui, le 20 décembre 1888, prescrivait par téléphone d'user de bienveillance à l'égard de M. Salis'.

51   Letter, 11 March 1887, Archives Nationales, F. 21 338. '[S]i l'on pouvait, sans trop d'inconvénients, user d'une certaine tolérance à l'égard des cabarets situés dans un milieu spécial d'artistes comme celui de Montmartre, il y avait de réels dangers à user de la même tolérance à l'égard de ceux qui sont fréquentés par la jeunesse des écoles'.

The cabarets of Montmartre, therefore, were explicitly marked out by the state as a space of freedom that could be tolerated only in a carefully defined area on the margins of the city.[52] In doing so, the state itself willingly participated, to a certain extent, in the cabarets' performances of marginality and separation. So whilst some of the artists and spectators who spent time in Montmartre perceived the area as a quasi-utopian space in which new forms of morality, politics, and cultural expression could be practised free of outside interference, the cultural politics of Montmartre should be seen in terms of wider state strategy of managing social, political and cultural dissent by confining it to specific areas that were seen to be less likely to adversely affect potentially rebellious youngsters.

In this respect, therefore, Montmartre's very marginality and distance from the centre of Paris made it a privileged space of biopolitical government through contingency in the early Third Republic, both by the state and by the artists and audiences who located themselves them there. The life of the city in the newly liberalized Republic was not to be controlled through the disciplinary grids and techniques seeking to define the precise nature of desirable behaviour, but through biopolitical mechanisms that could leave room for a proliferation of contingencies, surprises and petty disruptions. The experimental freedoms allowed to the district of Montmartre, that is, can be seen as experiments in the liberal art of governing *through* freedom.[53]

Montmartre emerged in the 1880s, that is to say, as a kind of social laboratory of conduct. And like a laboratory, the activities there were closely monitored and carefully recorded. If the authorities exercised tolerance with regard to many of Montmartre's excesses, the police archives reveal that they kept a scrupulously close eye on the activities there, and submitted numerous detailed reports of their observations. Although the reports rarely discussed the artistic performances in the cabarets (unless they contained seditious content), they gave detailed accounts of the numbers of people attending, their social background, and descriptions of the cabaret's decoration and staff. Reading through reports of the Chat Noir cabaret (which, until the 1890s, was by far the most closely observed café in Montmartre), three specific themes are particularly prominent in their reports.

First, police agents were careful to note any possible political activities in cabarets, such as private meetings, seditious songs, or political speeches. Until the flowering of the anarchist movement in the 1890s, however (discussed in Chapter 8), on the whole they found little to trouble them unduly. Performers at most Montmartre cabarets knew full well that they monitored (with undercover police spies recording being identified and gently mocked by the performers) and were careful to stay on the right side of the law. Seditious activities recorded by the police were very minor. On 14 June 1885, for example, a highly suspicious agent reported that an evening event illegally requiring letters of invitation had

---

52    Pillet, 'Cafés-Concerts et Cabarets'.

53    P. Joyce, *The Rule of Freedom: Liberalism and the Modern City* (2003), Nikolas Rose, *Powers of Freedom: Reframing Political Thought* (Cambridge, 1999).

been held in the Chat Noir.[54] A few days later, agents launched an investigation into allegations that monarchist slogans were being sung at the cabaret. The agent reported that Rodolphe Salis, the owner of the cabaret, had defended himself by insisting that he hadn't managed to identify who had come out with these shocking outbursts, and suggesting that he had been the target of jealousies and animosities of unknown persons who were seeking to provoke a scandal and have him shut down.[55] No further action was taken. In October 1884, similarly, agents were shocked by a song that was disrespectful towards President Jules Grévy and his family, and, two years before Grévy was forced to resign following a cash-for-medals scandal involving his son in law, presciently made hints about his financial improprieties:

> The song which the correspondent Bonap reported to contain insults against the President of the Republic does indeed exist, but it had not been possible to procure any printed examples. Nevertheless I have managed to obtain a copy, which I submit attached to the present report. From the investigation which I undertook, it turns out that this song, which has 'The Grévys' for its title, was composed at the Chat Noir cabaret, 22, Boulevard Rochechouard, and printed at Bloh, rue Bloue.[56]

**The Grévys**
Happiest of all people
Are father Grévy,
Mother Grévy,
Son Grévy, daughter Grévy
Each member of the family
Lives exactly to their liking
At the Grévys.
Father Grévy used to have just one billiards table,
Only one billiards table, one paltry billiards table.
Today he's as rich as a boyar [a Russian aristocrat]

---

54   Police report, 14 June 1885, archives de la Préfecture de la Police, Estaminet Le Chat Noir 81000–16.

55   Police report, 24 June 1885. Archives de la Préfecture de la Police, Estaminet Chat Noir, 81000–16.

56   Police report, 24 October 1884. Archives de la Préfecture de la Police, Estaminet Chat Noir, 81000–16. 'J'ai l'honneur de rendre compte à Monsieur le Chef de la Police municipale … que la chanson signalée par le correspondant Bonap, comme contenant des outrages à Monsieur le Président de la République, existe en effet, mais qu'il n'a pas été possible de s'en procurer d'exemplaires imprimés. On est parvenu toute fois à en obtenir une copie que je transmets annexé au présent rapport. De l'enquête à laquelle j'ai fait procéder, il résulte que cette chanson, qui aurait pour titre «Les Grévy», aurait été effectivement composée au Cabaret du Chat Noir, 22, Bd Rochechouard, et imprimée chez Bloh, rue Bloue'.

A true boyar, a true boyar.
He saw that our true happiness is
For him to buy another one,
Beginning again the five billion
That was taken from us by the Germans.
Mother Grévy used to have just one bidet,
Only one bidet. ... [etc.].[57]

Second, police reports highlighted frequent breaches of the peace. The unusual mixture of bohemians, prostitutes, and pimps sharing the same space as curious bourgeois revellers and students created an intoxicating yet often inflammatory atmosphere. Salis clearly faced up uphill struggle in keeping criminal elements out of the cabaret, and often summoned police officers to help him eject troublemakers. The most dramatic episode recorded by the police, however, which received widespread coverage in the press, occurred after Salis had ejected a pimp from the cabaret. When the pimp returned with 20 companions demanding entry, Salis refused to let them in, and a brawl broke out in the street. During the brawl, one of the Chat Noir waiters received a blow to the head and died some hours later with a fractured skull. In a judicial inquiry, it was decided that the mortal blow had come from none other than Salis himself, who, shortly after being knocked unconscious by a chair, and in a state of some confusion, had accidentally struck the waiter. Salis was prosecuted for involuntary homicide but, in another show of leniency to Montmartre figures, was not jailed, and the cabaret was allowed to remain open. Many newspapers hinted that in fact the pimps were responsible for the death, but had clubbed together to form a united front against Salis. Police agents merely pointed out that there was nothing surprising in the incident. One wrote:

---

57   Report 24 October 1884, archives de la Préfecture de la Police, 81000–16.
  'Des gens ravis
  C'est le père Grévy
  C'est la mère Grévy
  Grévy fils, Grévy fille
  Chaque membre de la famille
  A son gré vit
  Chez les Grévy.
  Le père Grévy n'avait qu'un billard
  Rien qu'un billard, un seul billard
  Riche aujourd'hui comme un boyard
  Un vrai boyard, un vrai boyard
  Il a vu quel bonheur est le nôtre
  Pour pouvoir s'en acheter un autre
  Reprendre sur les allemands juillards
  La revanche de nos cinq milliards.
  La mère Grévy
  N'avait qu'un bidet
  Rien qu'un bidet ... ' etc.

> The Chat Noir affair has nothing extraordinary about it. There have been twenty brawls in this establishment. And that's simply because anyone who turns up badly dressed is received in the most insolent manner. Salis is certainly not at all interesting.[58]

Third, police highlighted the bad feeling that the cabaret was creating amongst those living in the area, due to the noise, throngs of people, and disruptions to the traffic. A report on the 17 June 1884, when the cabaret at the height of its popularity, described the scene:

> The establishment the Chat Noir, on rue de Laval, attracted an enormous multitude of revellers and visitors all yesterday evening. The rooms of the ground floor were constantly occupied from eight o'clock to midnight and refilled themselves at every moment; all the curiosity was extremely over-excited! At eleven o'clock the doors had to be shut for a short time in order to stop the flow of visitors. Among them were numerous well-heeled people in good company, many of whom had come from far away, and hackney carriages and chauffeur-driven cars brought members of society . ... The staff were in the same outfits as the day before, and this clothing, moreover, can be considered as one the main things that attracts the public's curiosity in the Chat Noir.

All this commotion, the police agent reported, was causing significant irritation amongst the local population.

> The street was blocked by carriages; the crowds pressed in and for the whole evening traffic was impeded. It is undeniable that such a state of affairs, if allowed to continue, will only exacerbate the weariness and prejudice of the inhabitants of the area, as well as of the people whose leisure or business take them to rue Laval.[59]

---

58  Archives de la Préfecture de la Police, Estaminet Chat Noir, 81000–16. 16 December 1884. 'L'affaire du Chat Noir n'a rien d'extraordinaire. Il y a eu vingt fois de bagarres dans cet établissement. Et, cela, tout simplement parce que quiconque représente chez lui mal vêtu est reçu d'une façon des plus insolentes. Au certes, Salis n'est pas intéressant du tout'.

59  Archives de la Préfecture de la Police, Estaminet Chat Noir, 81000–16.

'L'établissement le Chat Noir, rue de Laval, a attiré pendant toute la soirée d'hier une affluence énorme de consommateurs et de visiteurs. Les salles du rez-de-chaussée ont été, de huit heures à minuit, constamment occupés par un public se renouvelant à tous instant et tous la curiosité était vivement surexcitée!

Parmi ceux-ci les gens bien mis et de bonne compagnie étaient nombreux beaucoup tenaient de loin et arrivaient au fiacre et des voitures de maitre ont amené des gens du monde, qui ne semblaient pas des moins empressés à jouir la coup d'œil que présentait l'établissement. ... Le personnel de la maison était dans la même tenue que la veille, et cette tenue, d'ailleurs, peut être considérée comme un des principaux éléments de la curiosité qui se manifeste dans le public à l'égard le Chat Noir.

The close police surveillance of the cabaret indicates that whilst the Montmartre cabarets were indeed afforded a far greater degree of tolerance than anywhere else in the city, nevertheless police were ready to intervene as soon as activities crossed certain thresholds of acceptability. The performances in Montmartre's cabarets, that is, could challenge certain bourgeois codes, conventions and customs, but police were ready to intervene at the moment that this crossed the limits of allowable behaviour. Nevertheless, Montmartre became a symbol of free speech, and whenever this was threatened, such as when the Pierrot Noir was shut down by police, supposedly for flouting safety regulations, furious protests were voiced in several major newspapers.[60] It was when this carefree cultural protest started to translate itself into organized political dissent in the 1890s, as I discuss in Chapter 8, that cabarets were quickly closed down.

## Conclusion

Understanding the cultural politics of humour in fin-de-siècle Montmartre, this chapter has argued, requires making some important theoretical moves. First, since the role of humour was not only to create new affective experiences of vitality but also to use this sense of vitality to problematize dominant discourses of biological life, it is not enough to simply dismiss such humour, as Adorno and Horkheimer might, as a way of excluding forms of 'other', undomesticated life. Indeed, as I will explore further in following chapters, Montmartre artists were aiming precisely to uncover new forms of urban vitality and liveliness. Through 'experimental embodiments' of psychological maladies, much Montmartre humour aimed to make the limits of life *visible* and *felt*, and hence to find ways of testing the limits of life.

In addition, the relative freedom from censorship and political interference that was granted to Montmartre's avant-garde urban culture needs to be conceptualized in terms of an experiment in urban government that was conducted simultaneously by Montmartre bohemians and government officials. Whilst Montmartre was given a freedom to challenge bourgeois morals and knowledges that was not granted to other areas of the city, this was part of a wider experiment in more liberal forms of urban government that attempted to govern by proliferating, rather than eliminating, difference and contingency. Montmartre, that is, was the privileged site in the experimental elaboration a new diagram of urban government.

---

La rue était obstrue par les voitures; la foule se pressait et pendant toute la soirée la circulation y a été des places difficiles.

Il est incontestable qu'un tel état de choses ne saurait de prolonger dans qu'il eut résultat ennui et préjudice pour les habitants du quartier, ainsi d'ailleurs que pour les gens que leurs alésoir ou leurs affaires peuvent appeler le sois rue de Laval'.

60    See, for example, 'La Chanson Traquéè. Fermeture Du Cabaret Du "Pierrot Noir"', *Le Matin*, 18 April 1897.

A key part of the attraction of the humour of Monmartre's cabarets, I have argued, was their attempt to test and transgress established discourses concerning the 'life of the city'. In the following three chapters, I wish to unpack this argument in more detail by exploring the ways in which Montmartre cabarets explored the nature of urban vitality through three connected modes of experience: *affect, representation* and *perception*. It is to the Montmartre's spatial ethics of affect that we now turn, via a detailed case study of the Chat Noir cabaret, the performance space which became the symbolic home of the Montmartre counter-culture.

# PART II
# The Chat Noir

Chapter 5
# Defiant Laughter:
# Humour and the Aesthetics of Place

If the Moulin Rouge dancehall became the symbol of the more commercial spirit of 1890s Montmartre, it was the Chat Noir cabaret that remains one of the most enduring symbols of the radical urban culture of the 1880s. In these three chapters I offer a detailed exploration of the urban culture, the forms of authority, and use of humour in the literature and performances at the Chat Noir. The three chapters tackle, respectively, the affective, representational, and perceptual life of the city in the cultural experiments of the Chat Noir. In this chapter, my aim is to build upon geographical analyses of the role of humour in the production of space and place by excavating the affective elements of the Chat Noir's urban ethos, analyzing the distinctive ways in which its creative experiments sought to create a new style of urban living and to engineer a novel affective experience of urban place.

The foundation of this ethos was an interest in creating a new, dynamic experience of place by means of humour, irony and buffoonery. Laughter had a particularly powerful place in the late nineteenth-century urban politics of affect, and was a means by which the tragic elements of modern life could be re-appropriated in creative and productive new ways. Through humour, Montmartre could be transformed from a deprived, lifeless neighbourhood, far removed from the supposed benefits of modernization, into a lively, dynamic neighbourhood possessing a powerfully felt sense of place. In seeking to understand this, the chapter contributes to an understanding of the social role of humour in producing space, place and landscape. In humour, as we will see, the Montmartre avant-garde found a way of bringing new life to the urban environment and experimenting with new ways of creating an experience of place.

## The Free and Proud City of Montmartre

In 1884 a poster for the municipal elections could be seen displayed throughout Montmartre. 'The day has finally come', it announced,

> when Montmartre can and must claim its rights of autonomy against the rest of Paris ... Montmartre is wealthy enough in finance, art and spirit to lead its own life. Electors! This is no mistake! Let the noble flag of Montmartre flutter

in the winds of independence ... Montmartre deserves to be more than an administrative ward. It must be a free and proud city.[1]

The candidate in question was Rodolphe Salis, patron of the Chat Noir cabaret, which had opened some three years earlier.[2] His campaign promises were short and simple:

1. Separation of Montmartre from the state;
2. Nomination by the people of Montmartre of a Municipal Council and a Mayor of the New City;
3. Abolition of the local area tax and replacement of this hurtful levy with a tax on the lottery, reorganized under the direction of Montmartre, which would allow our district to meet its needs and to help the nineteen mercantile, miserable districts of Paris.
4. Protection of public food. Protection of workers nationwide.[3]

Salis' campaign was as much a publicity stunt as a serious political campaign – his campaign slogan was 'Serious as ever' – and he missed election. One thing that the poster reveals, however, is a fierce attachment to the neighbourhood of Montmartre. Montmartre was not just a place, but the expression of an ethos, a style of life. It had become an emblem of autonomy, freedom and creativity. And it was this intensely felt sense of place that made Montmartre such an attractive neighbourhood for a cultural and political avant-garde who saw Montmartre as 'an alternative society where creativity would be rewarded and eccentricity tolerated, a high-spirited realm where art rather than lucre determined status and social relations'.[4] Montmartre had come to symbolize

---

1   Émile Goudeau, *Dix Ans de Bohème* (Paris, 1888), pp. 274–5. 'Le jour est enfin venu où Montmartre peut et doit revendiquer ses droits d'autonomie contre le restant de Paris ... Montmartre est assez riche de finances, d'art et d'esprit pour vivre de sa vie propre. Électeurs! Il n'y a pas d'erreur! Faisons claquer au vent de l'indépendance le noble drapeau de Montmartre ... Montmartre mérite d'être mieux qu'un arrondissement. Il droit être une cité libre et fière'.

2   See Phillip Dennis Cate and Mary Shaw (eds), *The Spirit of Montmartre: Cabarets, Humor, and the Avant-Garde, 1875–1905* (New Brunswick, NJ, 1996), Louis Chevalier, *Montmartre Du Plaisir et du Crime* (Paris, 1995), Mariel Oberthur, *Le Cabaret du Chat Noir a Montmartre* (Geneva, 2007), Gabriel Weisberg (ed.), *Montmartre and the Making of Mass Culture* (New Brunswick, NJ, 2001).

3   Goudeau, *Dix Ans de Bohème*, p. 275. '1. La séparation de Montmartre et de l'État; 2. La nomination par les Montmartrois d'un Conseil Municipal et d'un Maire de la Cité Nouvelle; 3. L'abolition de l'octroi pour l'arrondissement, et le remplacement de cette taxe vexatoire par un impôt sur la Loterie, réorganisée sous la régie de Montmartre, qui permettrait à notre quartier de subvenir à ses besoins et d'aider les dix-neuf arrondissements mercantiles ou misérables de Paris; 4. La protection de l'alimentation publique. La protection des ouvriers nationaux'.

4   Richard Sonn, *Anarchism and Cultural Politics in Fin-De-Siècle France* (Lincoln, NE, 1989), p. 94.

an ideal that in many ways embodied the anarchist version of utopia, not only in its championing of free creativity or local autonomy, but also in its balancing of the rural and the urban elements, the gardens and the cabarets ... it preserved its own sacred space from which to gaze down upon the metropolis, countering its economic dependence with cultural autonomy and radicalism.[5]

What was special about Montmartre, then, was its vigorous assertion of a distinctive place-identity and its fierce protection of local autonomy. The Chat Noir, and the artistic groups attached to it such as the Fumistes, the Hydropaths and the Incoherents, was at the heart of attempts to re-imagine the possibilities of urban community and to engineer a deeply felt experience of place. Their use of humour in order to achieve this sense of place is particularly striking.

## Romanticist Montmartre and the Destruction of Place

For the Romantics of the mid-nineteenth century, the village of Montmartre, not yet annexed by Paris, had been imagined as a rural idyll: an authentic refuge from the alienation of the Metropolis.[6] Gérard de Nerval wrote in 1855:

> I lived for a long time in Montmartre, where one can enjoy very pure air, varied prospects and magnificent views ... There are windmills, cabarets and arbours, rustic paradises and quiet lanes, bordered with cottages, barns and bushy gardens, green fields ending in cliffs where springs filter through the clay, gradually cutting off certain small islands of green where goats frisk and browse on the thistles that grow out of the rocks; proud, surefooted little girls watch over them, playing amongst themselves.[7]

These comments evoke a typically Romanticist nostalgia for a purity of place and environment, isolated from the corrupting influence of urban industrialism. The experience of place, here, is tied to the benevolent circulation of nature's life and creativity. In Nerval's description, Montmartre is an area, overflowing with natural vitality, where life can be experienced in a more direct and pure fashion than is possible in the metropolis below.

---

5   Ibid., 94.
6   Philippe Jullian, *Montmartre*, trans. Anne Carter (Oxford, 1977).
7   Gerard de Nerval, *Promenades et Souvenirs. Lettres a Jenny Pandora* (Paris, 1972). Cited in Jullian. 'J'ai longtemps habité Montmartre; on y jouit d'un air très pur, de perspectives variées, et l'on y découvre des horizons magnifiques ... Il y a là des moulins, des cabarets et des tonnelles, des élysées champêtres et des ruelles silencieuses, bordées de chaumières, de granges et de jardins touffus, des plaines vertes coupées de précipices, où les sources filtrent dans la glaise, détachant peu à peu certains îlots de verdure où s'ébattent des chèvres, qui broutent l'acanthe suspendue aux rochers; des petites filles à l'oeil fier, au pied montagnard, les surveillent en jouant entre elles'.

The processes of modernization and urban expansion, however, quickly led Montmartre to acquire less idyllic characteristics. The area, largely proletarian, was formally annexed into the city in 1860. Émile Zola's 1877 description of lower Montmartre in *L'Assommoir* portrays it in terms of a natural world that is now irredeemably corrupted. Zola describes 'the drab slope of Montmartre amid the tall forest of factory chimneys streaking the horizon, in that chalky, desolate city outskirt'.[8] The process of modernization, in Zola's account, had destroyed the natural life and vitality of the neighbourhood. By the 1880s, Montmartre, through its association with poverty and the insurrection of the Paris Commune, was widely considered an unhealthy milieu in which natural life had become dangerously excessive or perverted. It seemed to be isolated from the healthy economic flows of the modern city – a neighbourhood in which any genuine sense of place or natural vitality had been destroyed.

Nonetheless, owing to cheap rents and remnants of a village-like feel, during the 1870s Montmartre had become a fashionable area for Impressionist artists to work. The art world was undergoing a rapid change during this time, as the collapse of the Second Empire in 1870 precipitated the end of the close relationship between artist and state. In 1880 control of the annual salon was relinquished by the government and taken over by the Société des Artistes Françaises. In the years following this, the systems of patronage were further undermined by the emergence of increasing numbers of commercial alternatives such as galleries, illustrated journals, and the new 'cabarets artistiques', which supported and provided publicity and funds for emerging artists.[9] By the 1880s, artists were no longer protected by the state, and found themselves fully exposed to the market. For all the insecurities that this created, it did allow them a new freedom to criticize both the state and the established institutions of art upon which they had previously relied to survive, thereby providing the conditions for the emergence of the first artistic avant-gardes.[10] New groups such as the Hydropathes and the Incoherents set to this task with zeal. Eventually they found their home in the Chat Noir, which soon became the loudest public voice of the spirit of Montmartre. Through humour, these artists sought a means by which to regenerate the life and vitality of the neighbourhood.

---

8    Émile Zola, *L'Assommoir*, trans. Margaret Mauldon (Oxford, 1995), pp. 251–3.

9    Phillip Dennis Cate, 'The Spirit of Montmartre', in Phillip Dennis Cate and Mary Shaw (eds), *The Spirit of Montmartre: Cabarets, Humor and the Avant-Garde, 1875–1905* (New Brunswick, 1996), Patricia Mainardi, *The End of the Salon: Art and the State in the Early Third Republic* (Cambridge, 1993).

10   Cate and Shaw (eds), *The Spirit of Montmartre: Cabarets, Humor, and the Avant-Garde, 1875–1905*, J. Henderson, *The First Avant-Garde 1887–1894: Sources of the Modern French Theatre* (London, 1971), R. Shattuck, *The Banquet Years: The Origins of the Avant-Garde in France 1885–1918*, 2nd ed. (New York, 1968).

**The Holy City**

On 14 January 1882, the first edition of the Chat Noir's satirical house journal was printed and distributed. From the start it aggressively promoted a new, modern myth of Montmartre. Its opening article, titled 'Montmartre', posed a humorous challenge to the Church, the emblem of transcendent authority. 'It is high time to correct an error which has weighed down on more than sixty entire generations', the article proclaimed. 'The writing which we call holy – I don't really know why – has done nothing more, to put it politely, than make a mockery of the people'.[11] The authority of religion was a particularly pressing matter for the inhabitants of Montmartre, since the Sacré Coeur Basilica, a much-hated symbol of Paris's collective penance for the revolt of the Commune, was in the process of being constructed there.[12] The Sacré Coeur was supposed to expiate Montmartre's sins, purifying a moral environment that had become diseased and dangerous. The Basilica was part of an attempt by monarchist-sympathizing politicians to create a new identity for Montmartre as a place of sacred pilgrimage, thereby undermining its association with working-class revolt.[13] In explicitly setting itself against the Church, the article immediately contested this new religious identity. In opposition to the new spiritual outlook envisioned by the state, it sketched the first elements of a different urban imaginary, one involving a rejection of traditional authority, an experiment of new, creative forms of modernity, and a celebration of the creativity and authority of embodied experience.

The article evoked an interestingly ambiguous imaginative geography. As well as portraying it as a place of anti-clericalism and anti-traditionalism, it also depicted Montmartre as the exact centre of the new modern world. The Biblical tradition, the article claimed, had erroneously neglected to identify Montmartre as the original soil of humanity. Through a series of word plays leading from 'Mont Ararat', via 'Mont-m'arrête' to 'Montmartre', the article 'proved' that after the flood, Noah had in fact cast his anchor, not at Mount Ararat, but at Montmartre. The text's ironic standpoint is obvious: religious imagery is used in order to assert the centrality of an emphatically secular ethos, one that was to be given place by Montmartre. Drawing on Montmartre's historical association with the martyrdom of Saint Denis, Salis habitually referred to Montmartre ironically as 'the holy city', and in doing so, announced his rejection of religion and all forms of traditional authority. Montmartre was to be the centre of a new, anti-authoritarian world.

---

11    Jacques Lehardy, 'Montmartre', *Le Chat Noir*, 14 January 1882. 'Il est grand temps de rectifier une erreur qui a pesé sur plus de soixante générations complètes. L'Écriture que l'on dit sainte – je ne sais trop pourquoi, – n'a fait pour parler poliment, que se moquer du peuple'.

12    David Harvey, *Paris: Capital of Modernity* (New York, 2003).

13    Raymond Jonas, 'Sacred Tourism and Secular Pilgrimage: Montmartre and the Basilica of Sacré Coeur', in Gabriel Weisberg (ed.), *Montmartre and the Making of Mass Culture* (New Brunswick, NJ, 2001).

Ne bougeons plus ! Tout le monde y passera !

**Figure 5.1    'Nobody move!'. Illustration in *Le Chat Noir*, vol. 1,
14 January 1882**

'Thus Montmartre is the cradle of humanity … Montmartre is the centre of the
world … It is Montmartre where the first city of humanity was built'.[14]

In fact, Montmartre was to be more than a centre. The humour of the *Chat Noir*
journal portrayed it as a dynamic, centripetal force, drawing the rest of the world
inexorably towards it. An illustration in the journal shows a line of bourgeois
people, depicted as farmyard animals, waiting to be photographed outside the
Chat Noir (Figure 5.1). 'Nobody move!', the caption reads. 'Everyone must pass

---

14   Lehardy, 'Montmartre'. 'Montmartre est le berceau de l'humanité … Montmartre
est le centre du monde … C'est à Montmartre que fut construite la première ville de
la humanité'.

through here'.[15] The message was clear – Montmartre was to be the new centre of modernity. In this urban imaginary, Montmartre, at the same time as possessing an irresistible magnetic force, was to remain a singularity, fiercely autonomous from the rest of world. 'Montmartre is isolated because it is self-sufficient. This centre is absolutely autonomous. It is said that in a small village located a long way from Montmartre, which travellers call Paris, a local academy is discussing the conditions of municipal autonomy. It is a long time since this question was resolved in Montmartre'.[16]

Readers would hardly have needed reminding that 11 years earlier the issue of municipal autonomy had indeed been settled during the 'semaine sanglante' that brought the Commune to a devastating end.[17] The message of the article was clear, however. The suppression of the Commune spelled the end of one form of autonomy; but the Montmartre avant-garde was committed to finding a *new* solution to the problem of local autonomy. This ghost of the Commune was evoked again a few issues later, in a rumbustious article titled 'The Assault of Montmartre'.

> Today, April 1st, at a quarter past four in the morning, an attempted assault was launched against Montmartre. In defiance of the most solemn promises, Léon Gambetta, at the head of the troops of the Chausée-d'Antin, has infiltrated the country of Montmartre ... The call has sounded out on the mountain; the Moulin de la Galette has been put in a state of defence; the soldiers of the Sacré Coeur have been confined to their church ... The Montmartre homeland is not in danger; it is simply threatened. The henchmen of tyranny want to undermine our immemorial hills at their foundations. No! The strength, the muscular elasticity of the autochthons of the Mount of Martyrs are the guarantees of the future success of our armies.[18]

---

15  *Le Chat Noir*, 14 January 1882, p. 3. 'Ne bougeons plus! Tout le monde y passera!'

16  Lehardy, 'Montmartre'. 'Montmartre est isolé, parce qu'il se suffit a lui seul. Ce centre est absolument autonome. Dans une petite ville située a une grande distance de Montmartre et que les voyageurs nomment Paris, une académie locale discute, dit-on, les conditions de l'autonomie municipale. Il y a longtemps que cette question est résolue à Montmartre. Les peuples qui pullulent à la surface du globe n'ont qu'à venir s'en assurer'.

17  On the history of the Commune, see, for example, Robert Tombs, *The Paris Commune, 1871* (London, 1999).

18  A'Kempis, 'L'assaut De Montmartre', *Le Chat Noir*, 1 April 1882. 'Aujourd'hui, 1er avril, à quatre heures pour le quart du matin, une tentative d'assaut a été faite contre Montmartre. Au mépris des promesses les plus solennelles, Léon Gambetta, à la tête des troupes de la Chaussée-d'Antin, a pénétré dans le pays Montmartrais ... Le rappel a sonné sur la montagne; le Moulin de la Galette a été mis en état de défense; les soldats du Sacré-Cœur sont consignés dans leur église. ... La patrie Montmartre n'est pas en danger, elle est tout simplement menacée. Les suppôts de la tyrannie veulent saper dans leur base nos buttes

In an imaginary conversation with Léon Gambetta (the recently ousted Prime Minister, suspected by his adversaries of planning to engineer an executive dictatorship), who asks why Montmartre has revolted against Paris, a leader of the insurrection insists: 'Rebellion? You should know, ignorant traveller, that this is a just revindication of the rights of the first aboriginal people against its oppressor'. Montmartre's aim is 'to win independence, autochthony'.[19]

Four principle features are discernible in this humorous imaginative geography of Paris. Firstly, Montmartre was portrayed as neighbourhood in which traditional authority – both Church and State – was to be rejected and undermined at every opportunity. Secondly, it was a place that was to be at the centre of the modern world, the point through which it would be necessary to pass to be part of the most dynamic and subversive forces of modernity. Thirdly, Montmartre would also be a place that was autonomous from the rest of the world and, above all, from Paris. Even as the world passed through its gates, its borders would be fiercely protected. Finally, Montmartre's identity was to come from its claim to be the home to the cultural spaces that offered the most dynamic, vital and modern forms of urban experience.

This humorous representation of the spirit of Montmartre was highly ambiguous. How could it be, on the one hand, a place that required every true modern to pass through it and, on the other, a place that was fully isolated and autonomous from the outside world? The paradox is embodied by the frequent references in *Le Chat Noir* to the 'autochthons' of Montmartre. An autochthon is a 'son of the soil', a dweller with the deepest possible roots in the area that he lives in; so the idea of autochthony is one linking place to the richest possible depth of time and tradition. To be an autochthon is the very opposite of being an inhabitant of modernity, a celebrator of the fleeting moment. Here, however, the portrayals of Montmartre in *Le Chat Noir* proclaimed an intention, saturated with contradiction, to create a distinctively modern form of autochthony, a sense of place born of the most fleeting, unstable and iconoclastic impulses.

## The Vitality of Humour

'Place', construed in the broadest terms, can be defined as the ways in which geographic areas are experienced and made meaningful. It is usually considered

---

immémoriales. Non! La vigueur, l'élasticité des muscles des autochtones de la montagne des Martyrs sont les garants des futures succès de nos armes'.

19   Chanouard, 'Il Faut Lutter', *Le Chat Noir*, 8 April 1882. 'Moi (vivement). – Rébellion!!! Sachez, ô voyageur ignare que c'est la juste revendication du premier peuple aborigène contre l'oppresseur. De quel droit la plaine veut-elle gouverner la montagne? Gambetta. – Je ne discute pas, j'interroge. Quels sont vos projets? Moi. – Conquérir l'indépendance, l'autochtonie'. On Gambetta, see J .P. T. Bury, *Gambetta's Final Years: The Era of Difficulties, 1877–1882* (London, 1982).

to have three features: a geographic location; a material form; and an investment with meaning and value.[20] Since this meaningfulness is often achieved through tradition, authenticity and rootedness, place politics have occasionally been argued to involve an inherent conservatism, a resistance to change and progress.[21] It is important to recognize, however, that the ways in which the politics of place exceed reduction to competing discursive representations.[22] Place politics are centred upon the mobilization of affective experiences which can be far more contradictory and ambiguous than any specific representation of place. Understanding the ways in which place is created and contested, therefore, requires an analysis of the production of forms of experience which may not be reducible to any single, coherent representation of it. And humour, perhaps, is a powerful form of experience through which a distinctive sense of place can emerge – and one which has received little attention in the academic literature on place.

Humour – like place – often functions as a way of creating and consolidating an inside and outside. Scholars have discovered numerous ways in which humour has the power to include and exclude, since in order to work it relies upon nuanced sensitivities to shared histories, traditions or codes.[23] A shared sense of humour is a highly effective social bond, as well as means of naturalizing learned differences, and hence, social hierarchies.[24] As well as consolidating group boundaries and identities, however, humour can also disrupt them, and a variety of ways have been identified through which humour can challenge established representations and thereby operate as a form of creative resistance.[25] Discursive analysis of humour, then, has proved highly effective in uncovering the ways in which humour operates in the production, reproduction and contestation of hegemonic identities and representations. Academic scholarship on humour has paid less attention, however, to the non-representational aspects of humour: the means by which humour moves us in ways that are not reducible to their semantic content.[26] There is a powerful

---

20 Tim Cresswell, *Place: A Short Introduction* (Oxford, 2004), Thomas Gieryn, 'A Space for Place in Sociology', *Annual Review of Sociology* 26 (2000).

21 For example David Harvey, 'From Space to Place and Back Again', *Justice, Nature, and the Geography of Difference* (Oxford, 1996).

22 Nigel Thrift, 'Steps to an Ecology of Place', in Doreen Massey, John Allen and Philip Sarre (eds), *Human Geography Today* (Cambridge, 1999).

23 Sharon Lockyer and Michael Pickering, 'You Must Be Joking: The Sociological Critique of Humour and Comic Media', *Sociology Compass* 2(3) (2005), Sharon Lockyer and Michael Pickering (eds), *Beyond a Joke: The Limits of Humour* (Basingstoke, 2005).

24 Giselinde Kuipers, *Good Humor, Bad Taste: A Sociology of the Joke* (Berlin and New York, 2006).

25 For example. Simon Weaver, 'The "Other" Laughs Back: Humour and Resistance in Anti-Racist Comedy', *Sociology* 44(1) (2010). Marjolein t'Hart and Dennis Bos (eds), *Humour and Social Protest* (Cambridge, 2007).

26 For example Scott Sharpe, Maria Hynes and Robert Fagan, 'Beat Me, Whip Me, Spank Me, Just Make It Right Again: Beyond the Didactic Masochism of Global Resistance', *Fibreculture* 6 (2005).

corporeal, embodied, affective element to humour that needs to be understood more clearly. Understanding the affective qualities of humour requires investigating its dynamism and experiential intensity – the ways in which it mobilizes an experience of vitality and liveliness and draws authority from this vital experience.

Laughter is the expression of an affective energy: it literally moves us. Kant put this by saying that the gratification of laughter is a corporeal, 'animal' pleasure, one emerging from 'the furtherance of the vital bodily processes' and causing a 'feeling of health'.[27] During the nineteenth century, influential theorists such as Spencer, Schopenhauer, Nietzsche and Bergson came to regard humour as an expression of biological vitality. The English philosopher Alexander Bain, for example, argued that laughter is an expression of a physiological increase of vitality and a 'heightening of the powers of life'.[28] Laughter, Bain suggested, emerges out of a release from constraint, since such a release produces pleasure and an increase in nervous energy.[29] He saw in laughter an anarchic destructiveness, 'a rebellion against order – a temptation to a dangerous moment of anarchy against the severe demands of social constraint'.[30]

Such ideas were echoed throughout the second half of the nineteenth century. By the 1880s, laughter had been thoroughly incorporated into biopolitical discourses through which laughter became associated with an increase in life, health and vitality through non-representational forms of experience. As a mode of experience that was vital and close to nature, it was particularly associated with the primitive and the animalistic.[31] Charles Darwin had chosen laughter as the first demonstration of his thesis that humans share expressive gestures with animals such as monkeys, and that the reason for this is their descent from a common progenitor.[32] Laughter was also associated with hysteria, a much-debated fin-de-siècle psychological malady.[33] By taking up humour as their most characteristic form of expression, the Montmartre avant-garde were appropriating biopolitical discourses that problematized the body and its affects in relation to values such as health, vigour and liveliness. Through humour, place could be invested with new forms of affective vitality.

At the same time as participating in such biopolitical discourse, however, the Montmartre avant-garde set out to test its terms. Rather than accepting the

---

27   Immanuel Kant, *The Critique of Judgment*, trans. J. H. Bernard (New York, 2000).

28   Bain, cited in Michael Billig, *Laughter and Ridicule: Towards a Social Critique of Humour* (London, 2005), p. 95.

29   Alexander Bain, *The Emotions and the Will* (London, 1859).

30   Billig, *Laughter and Ridicule: Towards a Social Critique of Humour*, p. 98.

31   Rae Beth Gordon, *Dances with Darwin, 1875–1910: Vernacular Modernity in France* (Farnham, 2009), pp. 141–5.

32   Charles Darwin, *The Expression of the Emotions in Man and Animals* (London, 1872).

33   Martha Noel Evans, *Fits and Starts: A Genealogy of Hysteria in Modern France* (Ithaca, NY, 1991).

established republican preoccupation with promoting life, health, and organic urban growth, artists associated with the Chat Noir promoted a form of vitality that resulted from a *fall* from nature, a deliberate self-alienation from the organic life of the city.

## Irony and the Fall from Life

The Montmartre avant-garde, we have seen, drew upon a paradoxical affirmation of both autochthony and exile, both rootedness and mobility. In doing so, they adopted and transformed established Romanticist theorizations of nature, place and irony. One of the cardinal virtues in Romanticism was creativity. This partly motivated the Romantics' interest in the natural world, since nature, they believed, possessed a creative vitality that was absent from the mechanized and alienating world of the modern metropolis.[34] For many Romantics, therefore, the aim was to achieve as close a unity as possible with nature. For a subversive strain of Romanticism associated with the Jena school, however, the goal was not at all to achieve harmony with natural life.[35] For although nature was creative, it created according to its own innate tendencies. In contrast to this, humanity had the capacity to present itself as something other than its immanent self-development. Unlike natural life, human life was capable of being other than any fixed nature. 'The need to raise itself above humanity' wrote Schlegel, 'is humanity's prime characteristic'.[36] However, this also meant that human creation would always involve a fall from nature, a slowing of life into inert objectivity. This meant, in turn, that *irony* and *buffoonery* were fundamental features of the human condition: irony, because humanity is always something other than what it is; and buffoonery, since humanity is always falling from creative life into inert objectivity, like a clown slipping on a banana skin.[37] Through irony and buffoonery, human subjectivity distinguished itself from natural life and affirmed its distance from it as the condition of a specifically human – and therefore higher – form of creativity. It is the ethos of this strain of Romanticism, which emphasized mankind's separation from nature's immanent self-unfolding, which was taken up and reconfigured by the Montmartre avant-garde. By evoking an experience of place through ironic humour, they were at once asserting Montmartre's modernity, and also making a specific statement about their attitude towards it.

---

34  Onno Oerlemans, *Romanticism and the Materiality of Nature* (Toronto, 2002).

35  Philippe Lacoue-Labarthe and Jean-Luc Nancy, *The Literary Absolute: The Theory of Literature in German Romanticism*, trans. Philip Barnard and Cheryl Lester (Albany, 1988).

36  Friedrich von Schlegel, *Philosophical Fragments*, trans. Peter Firchow (Minneapolis, 1991), p. 96.

37  Claire Colebrook, *Irony* (London, 2004).

Irony is a difficult word to define; it is sometimes characterized as a figure of speech in which what is said is the opposite of what is meant. In such cases, irony implicitly affirms the stability of an underlying world of meaning secured by the context of a speaking subject. In nineteenth-century Romanticism, however, irony developed greater complexity, becoming not just an opposition between what is said and what is meant, but a way of saying one thing at the same time as allowing for the possible validity of its contrary.[38] Through irony, the world itself becomes dynamic, mobile, and contradictory. Indeed, the order between world and concept is reversed: it is no longer a case of life being interpreted through language and concepts, but a case of language and concepts being effects of an infinite, dynamic life force that goes beyond any context.[39] Such a life cannot be known, for knowledge works through static concepts. However, it can be *experienced*, and irony, with its ability to speak the contradiction by which the world both is and is not identical with itself, is particularly well suited to making the dynamism of the world perceptible in experience. Irony creates the conditions by which life, as a dynamic, creative force, becomes available to experience and knowledge.

Irony, then, can be seen as more than a mere figure of speech, but as a style of life – an ethos towards the modern world. By expressing something of modern life through ironic humour, it was possible to grasp an essential experience of modernity's mode of being – its contradictoriness, its lack of identity with itself. Through irony, it was possible both to describe the world as it was, but also the determination to transform it, to make it something other than what it was. Understood in these theoretical terms, it becomes apparent that the usefulness of irony for developing a new experience of place, for Montmartre bohemians, lay in its ability to allow the contradictions of modernity to become available to experience, rather being artificially or arbitrarily excluded, resolved, or sublimated. Through irony, a non-representational ethos towards the city could be explored through affective experience.

Interpreted in this way, the seriousness of intent behind the ironic stance of the Chat Noir's artists and writers becomes clear. Irony was a means by which a specific attitude towards the urban environment could achieve expression. First, irony was a way of making the fleeting, fugitive or contradictory elements of modern urban living perceptible in experience. Ironic humour offered a way to experience the dynamism and vitality of the city. Second, irony made it possible to develop an experience of place that incorporated contradictory elements: on the one hand, a celebration of modern cosmopolitanism, anti-traditionalism, and anti-authoritarianism; and on the other hand, acquiescence to the intensity and authority of experience associated with rootedness, autochthony, and autonomy. Ironic humour was a means of creating a new, affective experience of place in which an iconoclastic impulse to destroy tradition and autochthony could coexist

---

38   See Frederick Garber (ed.), *Romantic Irony* (Budapest, 1988).
39   Colebrook, *Irony*.

with a nostalgia for authenticity and rootedness. Through irony, authenticity could be found in creativity, and rootedness transfigured as an experience of exile.

Irony is a predominantly linguistic form of expression, a form of experience that is predicated upon humanity's separation from nature and its mastery of it through language. Its limits lie in its inability adequately to express the tragic element of modernity: the ways in which the accelerating dynamism and flux of the world evade individuals' control, and in which the forces of modernity threaten to make the individual a passive plaything of frighteningly impersonal, objective forces.[40] In order to find a response to this tragic vision of modernity, it was necessary to turn to more directly physical forms of humour.

## Pierrot and the Modern Carnivalesque

It was through the carnivalesque buffoonery of the traditional pantomime character Pierrot that the writers of the Chat Noir moved towards a more directly physical, corporeal, and libidinal experience of place. Pantomime was the art of dramatizing silence, and spoke to the body's convulsions, violences, and unrestrained appetites. The Montmartre avant-garde not only explored such forms of humour in its literature and performances, but also took it directly into the spaces of everyday life. In doing so, they experimented with a novel ethos towards the transformation of the affective economy of Montmartre.

In Adolphe Willette's cartoon depictions of Pierrot in the *Chat Noir* journal, pantomime was converted for the first time into comic strip form (Figure 5.2). In order to do this, Willette rejected the tradition by which cartoons were accompanied by explicative captions, leaving his drawings as 'histoires sans paroles' (stories without words). When readers complained that they could not understand the stories contained in these often ambiguous depictions, Willette replied: 'Comprenez l'écriture Pierrotglyphique!' – 'Understand Pierrot-glyphic writing!'[41] The spirit evoked by Pierrot involved a direct corporeality that would brook no linguistic mediation. It explored a life of sensation rather than representation.

Pierrot soon came to be recognized as the personification of Montmartre's urban imaginary. He was considered to stand for a whole community, persecuted but joyous. Pierrot was a timeless hero of French pantomime, and as a figure of the 'lowest' form of art, he offered fin-de-siècle bohemians an alternative identity, an ironic *disguise* for their creative ambitions. In a bourgeois world in which true

---

40   See Georg Simmel, 'The Concept and Tragedy of Culture', in David Frisby and Mike Featherstone (eds), *Simmel on Culture: Selected Writings* (London, 1997).

41   David Kunzle, 'The Voices of Silence: Willette, Steinlen and the Introduction of the Silent Strip in the *Chat Noir*, with a German Coda', in Robin Varnum and Christina Gibbons (eds), *The Language of Comics: Word and Image* (Jackson, 2001).

art and beauty went unrecognized, according to this way of thinking, such art had to disguise itself in base forms such as pantomime. This was not new as such. French Romanticism is full of situations in which a great hero is forced to adopt an ignoble disguise. Victor Hugo's Hernani was forced to become an outlaw because his legitimate claims to public recognition were denied; another of Hugo's heroes, Ruy Blas, strived to maintain honour in a valet's uniform; Musset's Lorenzaccio was a noble rebel who disguised himself as a pander and a cheap, often comic entertainer.[42] Baudelaire had described the hero of modernity as a secret agent, a traitor to his own class. The figure of Pierrot evolved through the century, however, and Willette was the first to depict Pierrot as a bohemian artist (and, conversely, the bohemian artist as Pierrot). In doing so, Pierrot's iconography changed: he became noticeably paler in face and started to dress in black. His pallor, resembling that of moonlight, emphasized that he was an outcast, a child of the Moon. His black clothes were a parody of those of the bourgeois class: he was to present the bourgeoisie at once with a degrading parody of itself and an image of its victim.[43]

Pierrot's behaviour in Willette's cartoons is exuberant, playful, violent, and lascivious. On occasion his adventures end in success, and he ends up in the embrace of a beautiful woman (Figure 5.3). More often, he is abused, ignored, and sacrificed to the bourgeois god of money (Figure 5.4). There was a clear spirit of Rabelaisian Carnival invoked by this form of humour. However, it is too quick a step to associate the humour of the Montmartre avant-garde, as Michael Wilson does, directly with the kind of Renaissance carnivalesque spirit evoked by Mikhail Bakhtin.[44] Certainly Montmartre writers did exhibit a fondness for Rabelais, whose literature epitomizes the anarchic spirit of Renaissance carnival. Yet the Montmartre carnivalesque was a distinctively modern one. It responded directly to the saturnine pessimism of two of the most important cultural movements of late nineteenth-century modernity: Naturalism and Symbolism. Naturalism (associated with the novels of Zola, and inspired by evolutionary theory) and Symbolism (an artistic movement exploring the hidden associations of words and images, associated with figures such as Mallarmé and Maeterlinck) were both informed by a pervasive pessimism, a sense that in modernity, life and natural vitality was becoming irredeemably corrupted, and the social body starting to degenerate.[45]

---

42   Louisa Jones, *Sad Clowns and Pale Pierrots: Literature and the Popular Comic Arts in 19th-Century France* (Lexington, KY, 1984), p. 12.

43   Ibid.

44   Mikhail Bakhtin, *Rabelais and His World*, trans. Hélène Iswolsky (Bloomington, IN, 1984), Michael Wilson, 'Portrait of the Artist as a Louis XIII Chair', in Gabriel Weisberg (ed.), *Montmartre and the Making of Mass Culture* (New Brunswick, NJ, 2001).

45   Joseph Carroll, *Evolution and Literary Theory* (Columbia, 1995), F. Deak, *Symbolist Theater: The Formation of an Avant-Garde* (Baltimore, 1993), John Lucas, 'From Naturalism to Symbolism', *Renaissance and Modern Studies* 21(1) (1977).

**Figure 5.2** Adolphe Willette, 'Pierrot Fumiste', illustration in
*Le Chat Noir*, 18 March 1882

**Figure 5.3    Adolphe Willette, 'Pierrot Amoureux', illustration in**
***Le Chat Noir*, 8 April 1882**

**Figure 5.4**  Adolphe Willette, 'Passage de Venus sur le soleil', illustration in *Le Chat Noir*, 9 December 1882

The emphasis on buffoonery and pantomime humour at the Chat Noir can be interpreted as a direct response to this pervasive fin-de-siècle pessimism. The use of carnivalesque humour was a way of breaking both with the optimistic rhetoric of republican political discourse and the saturnine pessimism of the Republic's Naturalist and Symbolist critics. Symbolists, inspired by Schopenhauer, pursued mystical forms of experience that accessed a realm beyond will, life, desire and the body. They strived to create a wholly abstract *space* of experience, one in which the destabilizing dynamism of time and the tragic element of modern existence could be put at bay. The writers of the Chat Noir, by contrast, by insisting on giving *place* to experience in the new creative quarter of Montmartre, asked whether it might not be possible to develop an ethos towards the modern city which could develop through, rather than against, the dynamic forces of desire and the body.

According to Baudelaire, pantomime humour is a form of humour that speaks directly to the body.[46] In it, true comic savagery can be found. This makes the essence of laughter – its diabolical element – readily apparent. 'Laughter', Baudelaire writes, 'is satanic: it is thus profoundly human'.[47] In this motif of redemption via a fall, Baudelaire added a distinctively modern sensibility to the thinking of carnivalesque humour. The Garden of Eden, Baudelaire observes, was devoid of laughter and tears. When mankind enjoys a unity with God and with Nature, laughter and tears are not possible: they exist only after mankind has fallen from innocence. The pantomime clown, his laughter veiling his tears, embodies both of these emblems of humanity's fall from life. Pantomime humour stages the modern alienation between man and natural life. This alienation, however – this fall from nature – is not something to be mourned. Despite being continually abused by fate, the characters in pantomime never cease to affirm their existence. Baudelaire describes characters who

> feel themselves forcibly projected into a new existence. They do not seem at all put out. They set about preparing for the great disasters and the tumultuous destiny which awaits them, like a man who spits on his hands and rubs them together before doing some heroic deed ... Every gesture, every cry, every look seems to be saying: 'The fairy has willed it, and our fate hurls us on'.[48]

The Montmartre avant-garde adopted and radicalized pantomime buffoonery, using it as a way of developing an affirmative ethos that remained alive to modernity's violences, injustices and degradations. Like Baudelaire, the artists of the Chat Noir identified themselves with pantomime characters such as Pierrot.

---

46   Charles Baudelaire, 'On the Essence of Laughter', in Jonathan Mayne (ed.), *The Painter of Modern Life, and Other Essays* (London, 1964), p. 157.

47   Ibid., 160.

48   Ibid.

However, whereas Baudelaire merely used the clown as a metaphor for the modern artist, this later generation stylized a way of *living* as such clown-artist hybrids. As artists, they were destined to suffer, but by becoming artist-clowns, they could find a way to transfigure a life of suffering and affirm it, just as the characters of the pantomime do. And Montmartre was to be the home in which such a life of buffoonery could be practised.

Figures associated with the Chat Noir, then, experimented with creating a new, dynamic experience of place through an ethos towards modernity based upon the affirmative stance of pantomime humour. Performances at the Chat Noir involved a continual succession of japes and practical jokes. One typically macabre hoax, for example, involved advertising the death of the owner of the cabaret Rodolphe Salis. When mourners arrived at the cabaret to join the funeral procession, they found the cabaret 'open due to a bereavement', and Salis, posing as his brother, welcoming them in to enjoy a drink.[49] Other writers and artists also carefully cultivated this pantomimic persona. The poet Gustave Kahn described Willette as a real life Pierrot character:

> Willette is … the very pavement of the Paris streets, come alive with all its *blague*, all its wit, lit up by tenderness, giving off smoky glimmers of passing political passions. Willette is patriotic, Willette is working-class. He will be generous, he will be cruel, he will be sympathetic, he will be hateful, according to the direction of the wind … There is in Willette a figure who will man the barricades for the fun of it.[50]

This impulse towards a typically avant-gardist confusion of art and life is also discernible in several masked balls, processions and festivals that were organized on the streets of Montmartre. One such occasion was an anarchic festival organized by Willette called the 'Fête de la Vache Enragée', rivalling an official Parisian festival on the same day, and dedicated to those inhabitants of Montmartre who were reduced to hunger and starvation, or, in the popular idiom, forced to 'manger une vache enragée', that is, 'eat an angry cow'.[51] It asserted an affinity between workers and bohemians, both poor but fiercely proud of the independence and anti-authoritarianism of their neighbourhood. As the poster for the festival claimed: 'Citizens of the Butte [Montmartre]. Noble children of the sacred hill. Artists and merchants, artisans and poets. Everyone, especially in Montmartre, knows what the vache enragée is'. Against the charge that the festival was simply a romanticizing celebration of poverty, one organizer of the festival insisted:

---

49    Goudeau, *Dix Ans de Bohème*.

50    Cited in Jones, *Sad Clowns and Pale Pierrots: Literature and the Popular Comic Arts in 19th-Century France*, p. 194.

51    Laurent Bihl, 'L'"Armée du Chahut": Les Deux Vachalcades de 1896 et 1897', *Sociétés & Représentations* 27 (2001), Venita Datta, 'A Bohemian Festival: La Fête De La Vache Enragée', *Journal of Contemporary History* 28 (1993).

**Figure 5.5    Henri de Toulouse-Lautrec, illustration in *La Vache Enrageé*, vol. 1, 1896**

'By dedicating a parade to the Vache Enragée, we are scoffing at misery. Our laughter is not a grimace of submission or of complicity but rather one of defiance'.[52] Henri de Toulouse-Lautrec, the painter who produced the most evocative images of Montmartre popular culture, created the front cover of a journal that was published in conjunction with the festival (Figure 5.5). The poster encapsulates the sense of place embodied by the festival: a joyous Pierrot accompanied by a masked beauty, leading a revolt against a terrified bourgeois who flees down the hill to the safety of Paris.

Through carnivalesque humour, then, drawn from the tradition of French pantomime, the Montmartre avant-garde experimented with ways of creating an experience of place that was rooted in a sense of exile, suffering and oppression, and yet at the same time was capable of transfiguring this suffering into an affirmative attitude of joy and autonomy. The intimate connection of such forms of humour with bodily desires and forces enabled them to conceive of a modern form of humour which incorporated a connection with the soil of Montmartre. This degraded, primitive form of humour was to be an experience that created

52   Cited in Datta, 'A Bohemian Festival: La Fête de la Vache Enragée'.

the form of dynamic autochthony that they were searching for – a deeply felt, embodied connection with a place in continual flux. The embodied, affective experience of laughter was to serve as a form of experiential authority through which to resist, with authority, dominant biopolitical knowledges concerning the true life and vitality of the city in modernity.

## Conclusion: Urban Affect and the Politics of Laughter

The writers and artists associated with the Chat Noir, I have been arguing, used humour as a way of re-imagining the possibilities of urban place and intervening in the affective economy of the city. The Montmartre avant-garde looked in humour for a means of creatively transforming a sterile, deprived urban neighbourhood, and developing a new, modern and dynamic experience of place. Through ironic humour and pantomime buffoonery, Montmartre artists attempted to invest the neighbourhood with renewed energy and vigour. Part of this affective energy was an experience of place which recognized modernity's ills at the same time as celebrating the possibility of an affirmative stance to the world that might transfigure life from something moribund and stifling into something joyful and creative. Through humour, as a way of organizing contradictory semantic meanings and a means of exploring corporeal, affective ways of inhabiting space, they attempted to creatively negotiate the contradictions of modernity and create an experience of place that was dynamic and full of vitality without obscuring modernity's arbitrary violence, gross brutality and accelerating inequalities.

In the avant-garde literature, art and performances at the Chat Noir, 'Montmartre' became the name for a certain *ethos*, a specific art of urban living, an attempt to bring art and everyday life together in new ways. It became the name for a certain kind of affirmative pessimism.[53] This ethos involved three principal components. First, it was an affirmation of varieties of experience which exceeded the type of cognition that operates according to a logic of either/or. It was a commitment to saying 'yes' to an experience that exceeds the necessity to judge, an experience in which difference, mobility and change could be invested with the weight and richness of autochthony. It was a commitment towards expanding contradiction rather than resolving them, and living these contradictions in the affective responses of the body. Second, it was a form of affirmative pessimism. It involved an acknowledgement of the tragic element of modern life, a recognition of the individual's powerlessness in the face of frighteningly impersonal modern forces. It recognized that power and domination would always exist, that desires would remain unrealized, and an unalienated unity with natural life was impossible and undesirable. It acknowledged humanity's fall from natural life, its necessary

---

53 In certain respects, their experiments arguably coincided with Nietzsche's philosophy of affirmative pessimism. See Joshua Dienstag, 'Nietzsche's Dionysian Pessimism', *The American Political Science Review* 95(4) (2001).

alienation, and searched for ways of making productive use of this fall, of creating new forms of life outside life. Third, it was an affirmation of the will in and of itself. Rather than judging desire according to whether it achieved satisfaction – an enterprise destined to lead only to pain and misery – this attitude involved a celebration of desire as an active, creative process in itself. Pierrot's continual failure to adapt to the demands of modernity, his continual suffering of violence, poverty and thwarted dreams, was met not with bitterness but with joy in the next adventure. Thus will and desire emerged as energies capable of expanding the limits and potential of urban life and vitality.

An alternative narrative is possible where the aesthetic construction of Montmartre as a space of artistic freedom and affective energy is viewed as a cynical exercise in place-marketing to lure an upper-middle class audience who were titillated by the possibility of safely observing the encountering the dangers of a genuine working-class neighbourhood, and who enjoyed being ridiculed and insulted in venues such as the Chat Noir in order to experience a safe and unthreatening expression of proletarian revolt.[54] At the Chat Noir cabaret, bourgeois audiences could be confident that this expression of revolt would never develop beyond throwing insults and jokes to manning the barricades. Its performances can be argued to have commodified and aestheticized the revolutionary spirit of the Commune, laying the ground for the area's transformation into a hub of the emerging culture industries and the eradication of the memory of Montmartre's association with the radical politics.

Such criticisms would not be unfair. What needs to be added to this critique of the inability of the Montmartre avant-garde of the 1880s to develop a practical politics of resistance to match their affective and symbolic practices of resistance is their intervention in a different economy to that of the market: the biopolitical urban economy of life, vitality and affective experience. The laughter of Montmartre bohemia was not just the humour of frivolity or submission; it was a *defiant* laughter, a way of performatively enacting, through the dynamics of corporeal experience, a transformation of the life of the neighbourhood, bringing new vitality and buzz into Montmartre's hitherto dead environment. In this respect, Montmartrois laughter acted as a way of intervening in the biopolitical economy of urban vitality and affect. Rather than adhering to the dominant republican framing of urban vitality in terms of organic health and free circulation, the Chat Noir experimented with alternative experiences of place based upon irony, buffoonery, and an ethos of affirmation. Such an experience of place would become extremely important in the twentieth century, both through the radical movements that

---

54    On this phenomenon in Victorian Britain, see Seth Koven, *Slumming: Sexual and Social Politics in Victorian London* (Princeton, 2004). On the relationship between bohemians and bourgeoisie, see Mary Gluck, *Popular Bohemia: Modernism and Urban Culture in Nineteenth-Century Paris* (Cambridge, MA, 2005), Jerrold Seigel, *Bohemian Paris: Culture, Politics, and the Boundaries of Bourgeois Life, 1830–1930* (New York, 1986).

invented new forms of political opposition through creativity and play, and through the neoliberal techniques that appropriated the same tools for radically different ends. The cultural politics of fin-de-siècle Montmartre marked a pivotal point at which experience became the final authority through which to make political claims concerning the true nature of the life of the city.

# Chapter 6
# Counter-display and the Exhibition of Error

As we have seen, one important element of the Chat Noir's attempts to use humour to lend authority to experiments with reimagining the life of the city was its ethos of affirmation: its emphasis on saying 'yes' to modern life despite all its cruelties and violences. Such forms of affirmation, however, required rejecting the dominant ways in which life was known and controlled through representation. A key focus of the Chat Noir, for this reason, was a critique of modern practices of urban representation and its tendency to exclude the untidy, excessive, or transgressive elements of metropolitan life. My intention in this chapter is to shift focus from the affective life of the city to its representational life, uncovering the ways in which Montmartre culture attempted to challenge the authority of modern modes of representation by exposing its failures, errors and hypocrisies, as well as extracting new forms of experiential life from these failures.

Nineteenth-century urban culture is routinely described in terms of the dominance of images, the eye and 'scopic regimes of modernity'.[1] Nineteenth century urba culture was preoccupied with questions of visibility and invisibility, depth and surface, and thus with the relays between what is hidden and what is open to view, between esoteric and exoteric knowledge and experience. As we will see, the Chat Noir set out to test the limits of the role of representation in securing a certain relationship between the visible and the invisible. And this involved a distinctive ethos towards the life of the city.

We have already seen that central to the Chat Noir was a creative culture of parody, satire, and something akin to what Gustave Lanson, in his influential history of French literature, termed 'l'esprit gauloise' – a culture of ribald, transgressive, anarchical humour.[2] Two forms of representation, in particular, emerged as prominent targets of parody. The first was the culture of museum display, which was becoming an ever more important governmental space during the Third Republic, and which created totalizing displays of historical and evolutionary progress, making visible the invisible forces of time and life. The second was the culture of imperialist display, exemplified by the growing taste in popular newspapers for heroic tales of exploration in savage and uncivilized lands. These travellers' narratives served to make life and time visible in a slightly

---

1　Jonathan Crary, *Techniques of the Observer: On Vision and Modernity in the Nineteenth Century* (Cambridge, MA, 1990), Martin Jay, 'Scopic Regimes of Modernity', *Force Fields: Between Intellectual History and Cultural Critique* (New York, 1993), David Michael Levin (ed.), *Modernity and the Hegemony of Vision* (Berkeley, CA, 1993).
2　Gustave Lanson, *Histoire de la Literature Francaise* (Paris, 1912).

different way, imposing strict distinctions between savage, unordered life and civilized, orderly life, organized through representational visual display. As I will show in what follows, the Chat Noir opposed these practices of making visible by putting into practice an ethos of 'counter-display' which exposed the failures of representation, and extracted new forms of vitality from these failures.

## Scattering Time

The Chat Noir cabaret captured a popular anti-bourgeois spirit that ridiculed the Republic's obsession with self-improvement, progress and positive knowledge. In contrast to the orthodox republican values of scientific progress, financial profit, and moral respectability, Montmartre bohemians adopted other values better associated with aristocratic or proletarian culture such as heroism, beauty, sensuality, laughter and revolt.[3] The Chat Noir valued, above all, pleasure, humour and *esprit* – intended as a stark contrast to dreary republican commerce. Bourgeois Paris was at once captivated and appalled by it. This ambivalence is nicely expressed in an article in *La Construction Moderne*, an establishment architectural journal, which celebrates the exuberance and playfulness of the Chat Noir at the same time as relegating it, in the final sentences, to a mere distraction from serious matters.

> The Chat Noir is the temple of architectural and decorative eclecticism; all
> styles are embraced in its décor ... the current trends in art and construction
> are more chatnoiresque than you think; the facades of our homes are dull,
> empty and terribly uniform and boring to see, and we find that it is necessary
> to enlighten them, rejuvenate them, with something unexpected, picturesque,
> bizarre even, which cannot be obtained through a concentrated style or strict
> compliance with an aesthetic rule. The Chat Noir has put into practice, without
> doubt, one of the most ardent aspirations of our epoch ... tomorrow the whole
> of France will understand that the broken line is less dull than the straight line,
> that liberty of architectural form must be in direct accord with the liberty of
> ourpolitical form.

This tongue-in-cheek celebration of the venue's architectural style, however, came only through a simultaneous devaluation of it: '... Forgive me, serious people, look at the thermometer, take into account the fact that I have been occupied for

---

3    Phillip Dennis Cate and Mary Shaw (eds), *The Spirit of Montmartre: Cabarets, Humor, and the Avant-Garde, 1875–1905* (New Brunswick, NJ, 1996), Jerrold Seigel, *Bohemian Paris: Culture, Politics, and the Boundaries of Bourgeois Life, 1830–1930* (New York, 1986), Gabriel Weisberg (ed.), *Montmartre and the Making of Mass Culture* (New Brunswick, NJ, 2001), Elizabeth Wilson, *Bohemians: The Glorious Outcasts* (London, 2000).

long weeks on the most difficult and serious architectural problems; ... consider also that the holiday month has arrived!'[4]

This ambiguous response to the Chat Noir clearly perceives it as a threat to established aesthetic values, a threat to be embraced only by allowing it space at the margins of aesthetic concerns, rather than allowing it directly to challenge established ways of arranging space. Humorous playfulness could be accommodated, as long as it did not interfere with serious matters. However, in the material arrangement of space at the Chat Noir, it is possible to discern a challenge to this bourgeois solemnity.[5]

One effect that Salis seems to have aimed for in the decoration of the cabaret was a sense of a displaced present. The cabaret was decorated as a representational space that announced the inadequacy of representation, thereby challenging discursive models that cemented individuals' place in time. He styled the Chat Noir as a kind of anti-museum, a parody of an urban institution that had acquired a privileged role in representing both the life of time and the time of life. In his decoration for the cabaret, Salis carefully arranged a heterogeneous collection of oddments and trinkets, some antique, some modern, some valuable, others valueless. The effect was a rich bric-a-brac style that was intended to contrast with anything contemporary and bourgeois. The furnishing comprised an odd assortment of dark oak antique chairs and tables, old copper pots, swords, coats of arms and paintings of artists associated with the cabaret. Emile Goudeau described it thus:

> A cat on the post; a cat on the window; wooden tables; massive, square, solid chairs (sometimes ballistics against attackers); huge nails, called the nails of the Cross; ... extended tapestries along the walls, above jewelled panels torn off old chests; ... a fireplace, which seemed destined never to be lit, as it sheltered under its mantle, and carried on its andirons, all kinds of trinkets: a bedpan, gleaming

---

4 Marcus du Seigneur, 'Vitruve et Gambrinus et Le Chat Noir', *La Construction Moderne* (1885), pp. 517–18. '[L]e Chat noir est le temple de l'éclectisme architectural et décoratif; tous les styles ont été conviés à son ornementation ... le mouvement actuel de l'art et de la construction est plus chat-noiresque que vous ne pensez les façades de nos demeures dont ternes, vides, effroyablement uniformes et ennuyeuses à voir, et nous trouvons qu'il faut les illuminer, les rajeunir, les ragaillardir par un je ne sais quoi d'imprévu, de pittoresque, de bizarre même, qui ne peut être obtenu par la concentration d'un style ou la stricte observation d'une règle esthétique. Le chat noir a mis en pratique, sans s'en douter, une des aspirations les plus ardentes de notre époque ... demain la France entière comprendra que la ligne brisée est moins maussade que la ligne droite, que la liberté de la forme architecturale doit être en raison directe de la liberté de la forme politique ... pardonnez-moi, ô gens graves, regardez le thermomètre, réfléchissez que je me suis occupé pendant de longues semaines, des questions archéologiques les plus sérieuses et les plus difficiles; ... pensez aussi que voici venir le mois des vacances!'

5 On the decoration of the Chat Noir, see also Michael Wilson, 'Portrait of the Artist as a Louis XIII Chair', in Gabriel Weisberg (ed.), *Montmartre and the Making of Mass Culture* (New Brunswick, NJ, 2001).

as if painted by Chardin; a genuine death mask (perhaps Louis XIII); giant pitchers – a jumble, but no firewood at all. On a corner of the counter, a bust, Unknown Woman, from the Louvre, and, above, a huge cat's head, surrounded by golden rays, as one sees in churches around the symbol of the trinity'.[6]

To observers, the decoration seemed mediaeval or Renaissance. In advertisements for the cabaret, Salis declared the style to belong to the age of Louis XIII:

LE CHAT NOIR
*Cabaret Louis XIII*
FOUNDED, IN 1114, BY A FUMISTE
84, Boulevard Rochechouart

Styling the cabaret as a mediaeval hostelry conjured up a disordered world of vagabonds, criminals and other low-lives, a spirit of carnivalesque excess of the kind celebrated in the writings of Villon and Rabelais. Referencing the pre-modern past was a way of expressing hostility to the cultural and aesthetic sensibilities of a prosaic age preoccupied with science, technology, and commerce.[7]

On closer scrutiny, however, this turn to the distant past was not simply a nostalgic celebration of pre-modern values. Rather, the arrangement of objects within the space of the Chat Noir appeared to scatter time altogether. Salis styled the cabaret as a temporal space in which elements of an ostentatiously inauthentic pre-modern past were superimposed upon other elements of a resolutely modern present. A pastiche of the past, a fragmented present, and a sensationally imagined future were juxtaposed in a single representational surface. A sign outside the Chat Noir proclaimed the cabaret to be the pinnacle of modernity, exhorting: 'Be modern!' On entering, however, the passer-by would discover that the height of the modern was not technological wizardry or the latest fashion, but a bizarre jumble of pasts, presents and futures. It seemed that in Montmartre, modernity was not to be equated with the living present. Within the walls of the Chat Noir, audiences were immersed within a self-consciously artificial historical temporality.

---

6   Émile Goudeau, *Dix Ans De Bohème* (Paris, 1888), pp. 255–6. 'Un chat en potence, un chat sur le vitrail, des tables de bois, des sièges carrés, massifs, solides (parfois balistes contre les agresseurs), d'énormes clous, appelés clous de la Passion … des tapisseries étendues le long des murs au-dessus de panneaux diamantés arrachés à de vieux bahuts (que Salis collectionnait dès sa tendre enfance), une cheminée haute, dont la destinée sembla plus tard être de ne s'allumer jamais car elle abrita sous son manteau et porta sur ses landiers, toute sorte de bibelots: une bassinoire, rutilante comme si Chardin l'eut peinte, une tête de mort authentique (Louis XIII peut-être), des pincettes gigantesques – un fouillis; mais de fagots, point. Sur un coin de comptoir, un buste, du Louvre, et, au-dessus, une énorme tête de chat, entourée de rayons dorés, comme on en voit dans les églises autour du triangle symbolique'.

7   Elizabeth Emery and L. Morowitz, *Consuming the Past: The Medieval Revival in Fin-De-Siècle France* (Aldershot, 2003).

LE CHAT NOIR

Le Coup d'État du 2 Novembre 1882

**Figure 6.1** **'Le Coup d'Etat du 2 Novembre 1882', illustration in** *Le Chat Noir*, **28 October 1882**

In fact, the effect was of a present in disguise as something else. Rather than making the present visible, as the point leading from a known past to a progressive future, the decoration of the cabaret effectively made the present disappear from view, dissolving into a chaos of disjointed objects.

An article in the *Chat Noir* journal exemplified this kind of dispersal of time. One article, on 28 October 1882 reported the sensational news of a coup-d'état, reminding the reader of Louis Napoléon's 1851 coup following the 1848 revolution (Figure 6.1).

> … Last night, the President of the Republic, yielding to the solicitations of his entourage, committed the ultimate crime: the violation of freedom! Without concern for honouring the past, violating the most holy and sacred commitments, distancing himself from the memory of '48, which his white hair ought to recall to him, the President has forfeited his honour. Imitating the conduct of Bonaparte, he has let himself be taken on a reactionary path ending in a coup d'état which will cover his name with shame and disgrace forever … The blood in the streets cries vengeance against gold in the coffers![8]

8 A'Kempis, 'Coup D'état Du 2 Novembre 1882', *Le Chat Noir*, 28 October 1882. 'Coup d'état du 2 Novembre 1882

La nuit dernière le Président de la République, cédant aux sollicitations de son entourage, a commis de dernier des crimes: violer la liberté!!! Sans souci d'un passé

The catch is that the article reports an event occurring on 2 November: five days *after* it was published. Temporal experience at the Chat Noir was represented as an experience in the present, but a present that could occupy multiple points in the past, present or future. What mattered was not its place in relation to a historical past or present, perhaps, but the nature of the *experience* to be undergone there.

### The Chat Noir as Anti-museum

Charles Rearick observes of the cabaret that 'the royal rustic interior was extraordinary enough that it could attract curiosity seekers as a museum draws tourists'.[9] Indeed, Salis in fact did everything he could to emphasize the cabaret's museum-like character, going so far as to publish a catalogue to guide visitors around the venue. The *Chat Noir Guide* took the visitor on a tour of the cabaret, describing it in terms that ostentatiously parodied the vocabularies of the bourgeois art museum. The book was a pastiche of the kind of museum catalogue with which middle-class visitors would have been easily familiar. In the *Guide* the writers painted a humorous portrait of the cabaret that expressed something of Montmartre's spirit and values, and ridiculed many of the core values and presuppositions concerning the role of culture in the life of the modern city. The *Guide* offered a very detailed, grandiloquent description of the cabaret's interior and exterior furnishings, drawing attention to the scattered 'exhibits' on display. It exaggerated wildly, inventing the provenance of most of the objects, and often associating them with celebrated figures from France's cultural past and present. A description of what Salis named the 'Salles de Gardes' (Guard Room), for example, captures the tone and style of the *Guide*, which took the pompousness and self-aggrandizing authority of the bourgeois museum catalogue to absurd proportions:

> What strikes visitor's eye from the start, as he enters the Great Guard Room of the Chat Noir Hostelry, is the monumental chimney. This Roman chimney, which carries at its summit the arms of the house of Salis – Saul on a field of blue assisted by a cat armed for war – is one of the most beautiful in France, and the model, illuminated by Grasset, was deposited in the National Library, on the orders of the Ministry of Fine Arts. Two heavy byzantine columns,

---

jusqu'alors intègre, au mépris des engagements les plus saints et sacrés, éloignant de lui jusqu'au souvenir de 48 que ses cheveux blancs auraient de lui rappeler, le président a forfait à l'honneur.

Imitant la conduite des Bonaparte, il s'est laissé emporter dans la voie réactionnaire jusqu'à commettre un coup d'Etat qui couvrira son nom d'opprobre et de honte à tout jamais ... Le sang dans les rues crie vengeance contre l'or dans les caisses!'

9   Charles Rearick, *Pleasures of the Belle Epoque: Entertainment & Festivity in Turn of the Century France* (New Haven, 1985), p. 59.

on top of which two hieratic cats gaze with large golden eyes at the Swiss Steps, support the entablature. Climbing on the missals, two other cats frolic on top of a coat of Genoese velvet, vainly searching for the moderns, to turn the others from their holy mission. But the two sacred cats, like two sphinxes, remain mute and immobile at their post, overseeing the proud motto of the nobility of Chatnoirville: Mount-joy – Montmartre![10]

Having briefly introduced each of the rooms, the *Guide* listed various objects, ornaments and paintings that were used to decorate the cabaret. Like a museum catalogue, the book outlined the objects' provenance, and furnished a description which would guide the aesthetic judgement of the reader. A feel for the typical character of the catalogue is given in entries such as these:

15. Small 15th century chest, originating from the collection of General Galliffet, offered to the Chat Noir by Monsieur the Marquis de Puyferrat.
16. Florentine dresser, offered to the Chat Noir by the Legation of Italy.
17. Chinese vase, brought from Nanjing by Monsieur Janvier de la Motte and offered to the Chat Noir by Monsieur W. Bouguereau.
18. Large Norman chest previously belonging to Monsieur Barbey d'Aurevilly; offered to the Chat Noir by Captain Travers of Mautravers.
19. 16th century lantern, originating from the 'Piot qui chante' cabaret, once frequented by the master François Villon. This superb windowed lantern was bought in 1846 by Monsieur Champfleury, who kept it until 1885, the date when it was offered to the Chat Noir by King Oscar of Sweden.[11]

---

10   Georges Auriol, *Le Chat Noir – Guide* (Paris, 1887), p. 9. 'Ce qui frappe tout d'abord l'œil du visiteur, lors qu'il pénètre dans la Grande Salle des Gardes de l'Hostellerie du Chat Noir, c'est sa cheminée monumentale. Cette cheminée romane, qui porte à son sommet les armes de la maison de Salis – Saule sur champ d'azur servi par un chat armé en guerre, – est un des plus belles qui soient en France, el la modèle, enluminé par Grasset, en a été déposé à la Bibliothèque Nationale, sur les ordres du ministre des Beaux-Arts. Deux lourdes colonnes byzantines, surmontées de deux chats hiératiques qui dardent sur le Perron des Suisses leurs larges prunelles d'or, en soutiennent l'entablement. Grimpés sur des missels, deux autres chats s'ébattent au-dessus du manteau de velours de Gênes, cherchant, mais en vain, les modernes, à détourner les autres de leur mission sainte. Mais les deux chats sacrés, ainsi que deux sphinx, demeurent immobiles et muets à leur poste, veillant sans cesse sur la fière devise des sires de Chatnoirville: Montjoye-Montmartre!'
11   Ibid., 12–13.
15.   Petit Bahut du XVè siècle, provenant de la collection du général de Galliffet, offert au Chat Noir par M. le marquis de Puyferrat.
16.   Dressoir florentin, offert au Chat Noir par la Légation d'Italie.
17.   Vase de Chine, rapporté de Nankin par M. Janvier de la motte et offert au Chat Noir par M. W. Bouguereau.
18.   Grand Bahut normand ayant appartenu à M. Barbey d'Aurevilly; offert au Chat Noir par le capitaine Travers de Mautravers.

**Figure 6.2**     **Adolphe Willette, *Parce Domine*, c. 1884. The title comes from a penitential Gregorian chant: *Parce Domine, parce populo tuo: ne in aeternum irascaris nobis* (Spare, O Lord, spare Your people: lest You be angry with us forever.) The hymn finishes: 'O beloved Searcher of Hearts, You know the weakness of mortal bodies; show to those returning to You the grace of forgiveness'**

Such entries made outlandish claims, both for the provenance and quality of the trinkets on display, and also for the social status of the cabaret's patrons. Ironically portraying itself as a temple of the arts, the *Guide* parodied the bourgeois pretensions of the art museum to act as the arbiter of cultural authority, taste and expertise.

Pride of place in the catalogue was given to a painting by Adolphe Willette (Figure 6.2). The huge painting covered one wall of the cabaret and depicted a carnivalesque torrent of bodies pouring down from Montmartre into Paris. It is not clear if the bodies are ghosts or living: a suicidal Pierrot leads the procession; a young woman rides on the back of an enormous black cat. A skull supervises the procession in place of the Moon, and angels dance in the sky. Montmartre seems to be poised at the intersection of the living and the dead. The painting captures the combination of excessive energy and macabre pessimism of the spirit of Montmartre. The catalogue describes the painting thus:

1.  Parce Domine, superb canvas by Willette (5m by 3m). It is, in the opinion of all amateurs and artists, Willette's masterpiece. It unites all the grace and

19.   Lanterne du XVIè siècle, provenant du cabaret du Piot qui chante, autrefois fréquenté par maître Françoys Villon. Cette superbe lanterne fenêtrées avait été achetée en 1846 par M. Champfleury, qui la conserva jusqu'en 1885, date à laquelle elle fut offerte au Chat Noir par le rois Oscar de Suède'

poetry which he has planted in so many exquisite compositions. To describe this painting would be a crime; before a canvas like this the emotions are so varied that it is impossible to record them. We can only say this: Come and see it. Or rather, come and see it again; for there isn't a man of spirit in all of Paris who hasn't already come to admire this vertiginous poem. Ordered by the Grand Duke Alexis, *Parce Domine* was offered to the Chat Noir in 1884, at the request of the tsar.[12]

What is immediately clear from the *Guide* is its light-hearted challenge to the austere bourgeois equation of 'culture' with 'civilization' – and the 'cult of secular, progressive self-development' which this idea of culture entailed.[13] The nineteenth-century museum was an exemplary space in which culture was used for education, self-improvement and moral reform. By parodying this kind of repository of official culture, the cabaret rhetorically positioned Montmartre as the seat of a very different form of culture, one which was not best encountered through dry representational techniques such as a catalogue, but as something to be *directly experienced*. Rather than offering a representational narrative of historical progress, it would offer a more direct experience of a present, a present which, defined neither by its relation to the past nor to its future, would be definitively modern. In order to understand how it achieved this, it is necessary to examine the wider politics of museum display in late nineteenth-century France.

## Displaying Error

The Chat Noir's anti-museum emerged during a period in which the museum was becoming a major urban institution of public instruction. The museum spoke to the new Republic's renewed emphasis upon reason, education, and progress. New city museums were built in many of France's provincial cities, and a new kind of museum, the ethnographic museum, came into being. These museums were not neutral spaces for the display of knowledge and cultural riches. Rather, they formed an important part of the late nineteenth-century reform of behaviours and morality. One way in which they achieved this was by using representational techniques to make life – biological life, urban life, and temporal life – visible in specific ways.

---

12   'Parce Domine, superbe toile de Willette (5 mètres sur 3). Le Parce Domine est, de l'avis de tous les amateurs et de tous les artistes, l'œuvre maîtresse de Willette. Il y a réuni toute la grâce et toute la poésie qu'il a semées dans tant d'exquises compositions. Décrire ce tableau serait un crime, devant une page comme celle-là les émotions sont si diverses, qu'il n'est pas à Paris un homme d'esprit qui n'ait admiré ce vertigineux poème. Commandé par S. A. I. le grand-duc Alexis, le Parce Domine fut offert au Chat Noir, en 1884, sur la demande du tzar'.

13   Terry Eagleton, *The Idea of Culture* (Oxford, 2000), p. 9.

The origins of modern museum display can be traced back to the 1789 revolution.[14] On 10 August 1792, the Tuileries Palace was overrun and Louis XVI was captured, imprisoned, and eventually sentenced to death. Just nine days after the king's capture, an official decree was passed that asserted the importance of collecting artworks and displaying them to the public in the former royal palace. Republicans saw this public display of culture to be a symbol of revolutionary egalitarianism and of the cultural benefits of liberty. Following the state's appropriation of property of the church and aristocracy (and later from foreign territories), it acquired a vast national patrimony. The new Louvre was to display these cultural riches in the hope of making Paris the centre of the world's cultural life, the Athens of the modern world. In 1793, the new museum was opened to the public, developing an approach to the cultural life of the population that would expand dramatically during later decades of the nineteenth century.

What was specifically modern about the new Louvre was, first of all, its democratic gesture of opening its doors to the general public, rather than a privileged élite. Another specifically modern feature was its innovative techniques for *displaying* its objects.[15] This was markedly different to royal and princely collections, which displayed sovereign power through an abundant arrangement of paintings whose effect was designed to be dazzling and overwhelming, and which made attention to individual paintings virtually impossible. In the displays of the Louvre a new system of display emerged, one which made visible historical evolution within national schools. This system explicitly referred to the method of classifying plants and animals by genus and species that was introduced by Linnaeus and Buffon in the mid-1700s. By creating an equivalence between cultural life and natural life, and classifying artworks in terms of their place in a linear narrative of evolutionary progress, the new Republic could displace the politically problematic iconographic content of the pictures (much of which depicted religious or aristocratic themes), and incorporate them into a form of temporal life governed by reason and natural order.[16] This new mode of representation testified to a growing belief that history followed an organic evolutionary path, and that by displaying it in terms of evolutionary progress, culture could be rationally classified and made to serve a useful purpose through well-ordered public exhibitions.

The modernity of the new forms of exhibition, therefore, lay in their use of representational classificatory devices in order to present culture as a form of organic life. This form of representation was adopted by the many public museums that were created in provincial cities during the early decades of the nineteenth century. By the 1880s, as greater interest was taken into the governmental possibilities of the museum, the museum acquired a prestigious role in the urban environment. This was due in large part to the interest in using education as means of creating progressive

---

14 Andrew McClellan, *Inventing the Louvre: Art, Politics, and the Origins of the Modern Museum in Eighteenth-Century Paris* (Cambridge, 1994).

15 Ibid.

16 Ibid., 80–81.

republican citizens.[17] The influential political theorist Émile Littré argued that education was an essential tool for strengthening the hold of the government upon the people, and undermining the grip of both Catholic ideas, at one extreme, and socialist ideas, at the other.[18] Education was a means by which to mould the nation's citizens and inculcate in them a desirable model of modern citizenship. Only properly educated individuals would become capable of participating fully in the life of the nation and the historical passage of civilization and progress, and the museum was a crucial means of extending education into adulthood.

One important function of museums was their disciplinary effect – the ways in which they acted as spaces in which the visitors themselves were put on display and learnt to reform their own morals and behaviours.[19] Here, however, it is the museum's mode of representation that is of most interest. The museum not only made life visible as something ordered, coherent, and with a clear goal; it also used representational means to promote a specific form of knowledge. The nineteenth-century 'universal survey' museum promised the visitor a direct encounter with reality.[20] In the museum, reality would be allowed to speak for itself in a way it could not in the chaotic spaces of everyday life. The method by which it could be made to speak for itself was a principle of representativeness. The museum would create a display of total knowledge, leading the visitor along a historical path on which representative samples of every major step in the evolutionary process (whether of art, history or nature) would be displayed. In this kind of museum display, the whole of history could be taken in at once. A museum visitor would rehearse a performance where knowledge of natural or cultural life would be gained through encountering representative samples of life as a totality. Museums displayed knowledge as a finished product, hiding the processes that created it, along with its tensions, contradictions and ambiguities. The museum was a space in which reality would not just be described to the visitor, but visually *presented* to them, supposedly offering the visitor a direct, first-hand experience of reality. The work of curatorship, therefore, had to be made invisible, so that the experience of the visitor could appear pure and unmediated, rather than artificially controlled and constructed by institutionally sanctioned experts.

The nineteenth-century museum operated through an aesthetics of truth which privileged direct visual presentation according to a principle of representativeness.

---

17   Kay Chadwick, 'Education in Secular France: (Re)Defining *Laïcité*', *Modern & Contemporary France* 5(1) (1997), J. Lehning, *To Be a Citizen: The Political Culture of the Early French Third Republic* (2001).

18   John Scott, *Republican Ideas and the Liberal Tradition in France, 1870–1914* (New York, 1951), p. 103.

19   Tony Bennett, *Birth of the Museum: History, Theory, Politics* (London and New York, 1995), Douglas Crimp, 'On the Museum's Ruins', *October* 13 (1980), Eilean Hooper-Greenhill, *Museums and the Shaping of Knowledge* (New York, 1992).

20   See C. Duncan and A. Walloch, 'The Universal Survey Museum', *Art History* 3 (1980).

This was a way of 'making visible' that relied on rendering invisible the relations of power, authority and expertise involved in the production of knowledge. The anti-museum at the Chat Noir, by contrast, brought the work of curatorship to the forefront. Its scattering of time satirized the hubris of museums' pretensions towards organizing the whole of history in one coherent display. By pretending to exert a greater control over time than even the bourgeois museum would lay claim to, the plurality of ways in which the past inevitably escapes any given representation of it could become apparent. Moreover, by creating a museum catalogue which was obviously and elaborately erroneous, they sent up the claims of the 'universal survey' museum to offer an objective and unmediated encounter with reality. In other words, they drew attention to the ways in which museum representations necessarily mediated the relation between object and viewer, and the possibility that these relations might not be disinterested. By drawing upon representations that were obviously filled with errors or falsehoods, the Chat Noir 'museum' mocked bourgeois claims to the authority to define the truth.

Rather than hiding the processes behind the production of representations, then, the Chat Noir put the processes embedded within the production of representation at the forefront. Experience, at the Chat Noir, was not a direct experience leading to a privileged knowledge of reality, but an indirect, mediated experience capable of disrupting established distributions of knowledge and experience. The Chat Noir's forms of display evoked forms of experience that were not safely ordered through representational codes, but which, through error and artifice, escaped this representational matrix. They gestured towards a form of life which, rather than being individuated through orderly representations, came into being as a form of error, a wandering away from established knowledge into different modalities of experience.

This emphasis on error was equally apparent in a different set of literatures devoted to giving an account of the fin-de-siècle urban landscape. These were texts that, by parodying the conventional tropes of colonial travel writing, searched for new ways of affirming a new experience of the life of the city as a form of erring, a deviation from the established pathways towards truth.

## Colonizing Paris

Perhaps the most important stage for the display of the temporal life of modernity was the city itself. For this reason, Montmartre artists could not just retreat to a neighbourhood whose reputation as a rural refuge from the modern city still lingered on; they had to venture further into the metropolis. In a number of articles in *Le Chat Noir* journal, the writers created a series of humorous urban landscapes in which they adopted the posture of the colonial explorer, venturing forth from the advanced civilization of Montmartre into the primitive, savage land of Paris. The series of articles, titled 'Voyages de Découvertes' and 'Lettres d'un Explorateur', parodied the heroic arrogance of the modern explorer, casting their powerful gaze, not on overseas territories, but upon the city of Paris, the self-proclaimed seat of modernity, civilization and progress.

The first article is an explorer's account as he leaves the 'free city' of Montmartre:

> Having taken in the rue des Martyrs, the traveller's vermouth-cassis, I put on once again my otter-skin cape, and taking hold of my arms – my stick and my purse – I struck out at low speed (hardly three knots an hour) to the famous Lorette pagoda [i.e. the church Notre-Dame-de-Lorette]. I passed through the place which navigators all call Carrefour Drouot, and exclaiming to myself: 'All right!' I launched myself into the unknown. For the traveller there is truly a moment, at once cruel and sublime, when, after having embraced all his nearest and dearest, the Montmartrais finds himself lost in the immense. All right![21]

Paris is imagined here as virgin territory, encountered for the first time with a naïve gaze that is impervious to the auratic power of the city's monumental buildings and boulevards. In doing so, the traveller enters a city which had in recent decades become semantically over-burdened. The new urban architecture was designed according to a strict representational architectural code, serving to make the spaces of the city legible at a glance.[22] The new urban fabric, with its rigid straight lines and carefully architected vistas, conveyed an impression of uniformity, order and authority. Through its geometry, the hierarchical relations between different areas of the city were immediately visible, and allowed no room for interpretation or doubt. The built environment was invested with a rich layer of representational code, demarcating the exact role, function and status of every inch of public space. The result of this was that the entire city became a vast exhibition of power, wealth and desire, of endless fascination to those who were granted the leisure and status to enjoy it. Impressionist artists, for example, celebrated enthusiastically the buzzing vitality of the bourgeois spaces of the modern city, largely overlooking any spaces which did not testify to a capitalist utopia of leisure, display and consumption.[23]

---

21    A'Kempis, 'Voyages De Découvertes', *Le Chat Noir*, 14 January 1882.

'Après avoir absorbé rue des Martyrs, le vermouth-cassis du voyageur, j'ai revêtu ma pelisse ornée de loutre, et m'étant muni de mes armes: ma canne et mon porte-monnaie, j'ai atteint à petite vitesse (à peine trois nœuds à l'heure) la fameuse pagode de Lorette. J'ai relevé en passant la position que tous les navigateurs désignent sous le nom de Carrefour Drouot, et m'écriant: All right! je me suis lancé dans l'inconnu. Il y a vraiment pour le voyageur un moment d'émotion à la fois cruelle et sublime, quand, après avoir embrassé tous les siens et les siennes, le Montmartrais se sent perdu dans l'immensité. All right!'

22    François Loyer, *Paris Nineteenth Century: Architecture and Urbanism* (New York, 1988).

23    Albert Boime, *Art and the French Commune: Imagining Paris after War and Revolution* (Princeton, NJ, 1995), Timothy J. Clarke, *The Painting of Modern Life: Paris in the Art of Manet and His Followers* (New York, 1985).

Mary Pratt highlights three frequent tropes of imperial travel narrative: aestheticization (making aesthetic pleasure the sole value and significance of the journey); density of meaning (representing the landscape as rich in material and semantic substance); and the relation of mastery predicated between seer and seen.[24] The *Chat Noir* explorers made liberal use of each of these techniques. The humour of their accounts, however, emerges from the explorers' *failure* to translate this representational code into practical experience. In these imaginary landscapes, the passage between experience and knowledge continually goes awry, becoming diverted and leading to absurd new situations. Once in the alien territory of Paris, the explorers, throwing their orientalizing gaze upon the lofty monuments and institutions of the city, implicitly challenge the modernity of the emblematic sites of Parisian culture – its museums, opera houses, churches and boulevards. Most often, the 'civilized' explorers reveal their inability to decode the city, becoming hopelessly confused as to the identity and function of the different monuments. In one story, for example, the explorer confuses the Louvre art museum with the Ministry of Finances, which was housed in the Louvre.

> I approached a large monument with wide doors, large windows, and no style.
> I was assured that this building was called the Louvre and had served old
> painters who made paintings for the government there. I was curious to visit this
> establishment, on the door of which, by a quirk of bad taste, has been written:
> Finance. I found a kind of gallery where people were locked in cages that were
> decorated with tiny doors and marked 'guichet', like the money changers on
> the rue Orsel at home. Other individuals with blue, yellow, green or white
> papers appeared at the windows and took money. These must be the modern
> painters of the country's government, but when I saw various posters, covered
> with numbers, plastered on the walls, I had very little desire to see the works
> which this enormous building must conceal ... You can reassure our friends in
> Montmartre that there is not a painting here that can compete with theirs. These
> number tables lack any taste, and reveal in their creators an absolute lack of
> idealism and selflessness.[25]

---

24  Mary Louise Pratt, *Imperial Eyes: Travel Writing and Transculturation* (London, 1992), p. 204.

25  A'Kempis, 'Voyages De Découvertes (Suite)', *Le Chat Noir*, 4 February 1882.

'Je me suis rapproché d'un gros monument sans style, avec de grandes portes et de grandes fenêtres. On m'a assuré que cette bâtisse s'appelait le et avait servi a de vieux peintres qui y faisaient des tableaux pour le gouvernement. J'ai eu la curiosité de visiter cet établissement sur la porte duquel par une bizarrerie de mauvais gout on a écrit: Finances. J'ai trouvé une manière de galerie où des gens étaient enfermés dans des cages ornées de toutes petites portes sur lesquelles on lisait guichet comme chez le changeur de la rue d'Orsel, chez nous. D'autres individus munis de papiers bleus, jaunes, verts ou blancs se présentaient à ces guichets et touchaient de l'argent. Ce doivent être les peintres modernes du gouvernement du pays; mais quand j'ai vu placardés sur les murs des espèces d'affiches, couvertes de nombres, j'ai eu un très faibles désir de voir les productions informes que doit

One frequent narrative device in these urban landscapes was to portray élite cultural spaces – supposedly the great expressions of the life of the city – as spaces of death or decay. For example, coming across Charles Garnier's Paris Opéra, the crown of Haussmann's urban renovation, the explorer mistakes the building for a tomb – also associating it, to make matters worse, with two overtly anti-republican symbols, the monarchy and Germany.

> Then … I saw a very tall monument, gilded, with a polychromatic appearance, with funeral urns, crypts, rostral columns, groups of women dancing in an infernal and deadly circle, reproducing very exactly certain frescoes which we know in Montmartre, and signed by two ancestors, Albert Dürer and Holbein. This must be an immense sepulchre in which the mummies of this country's ancient kings are entombed. Confirming this opinion is the title of the monument: 'Academy nationale de musique'. It is easily understood that the priests here teach cantiques [religious songs] to young girls, and if necessary make the young men of the country into eunuchs, charging them with celebrating in high-pitched tones the merits of dead monarchs.[26]

One effect of these literary landscapes is to challenge the city's self-appointed role as the motor of global modernity. Travel writing was an important means by which the authority of colonial imaginaries of modernity were communicated and embedded.[27] The heroic explorer was typically represented as an all-seeing master, able to make sense of everything he encounters. Furthermore, such travel writing was not always confined to the colonies. Within the emerging social sciences, the western metropolis was frequently described in terms evoking exploration narratives.[28] The *Chat Noir*'s humorous landscapes of Paris, by contrast, adopted the posture of the imperial explorer as a way of undermining the claim of bourgeois

---

recéler cet énorme édifice … Vous pouvez rassurer nos amis de Montmartre, ce n'est pas encore cette peinture-là qui peut leur faire concurrence. Ces tableaux chiffrés sont d'un goût atroce et indiquent chez leurs auteurs un manque absolu d'idéal et de désintéressement'.

26  A'Kempis, 'Voyages De Découvertes', *Le Chat Noir*, 21 January 1882.

'Puis … j'aperçois un monument très haut, doré, d'allure polychrome, avec des urnes funéraires, des cryptes, des colonnes rostrales, des groups de femmes tournant une ronde infernale et nécrologique, reproduisant très exactement certaines fresques que nous connaissons à Montmartre, et signées de deux ancêtres, Albert Durer, et Holbein. Ce doit être un immense sépulcre où sont enfermés les momies des Rois Anciens de ces pays. Ce qui me confirme dans cette opinion, c'est le titre même du monument: Académie nationale de musique. On comprend aisément que des prêtres se trouvant là, apprennent des cantiques aux jeunes filles, et rendent au besoin eunuques les jeunes gens du pays, chargés de célébrer en notes aigues les mérites des monarques défunts'.

27  James Duncan and Derek Gregory, *Writes of Passage: Reading Travel Writing* (London, 1999), Pratt, *Imperial Eyes: Travel Writing and Transculturation*, Tim Youngs, *Travel Writing in the Nineteenth Century: Filling the Blank Spaces* (London, 2006).

28  Felix Driver, *Geography Militant: Cultures of Exploration and Empire* (Oxford, 2001).

Paris to be the pinnacle of modernity and progress; and in doing so they positioned Montmartre as an alternative seat of modernity. In addition, however, they undermined the epistemological formation that linked colonial forms of display to the temporal life of the modern city.

### Exhibiting the World-as-exhibition

As Timothy Mitchell observes, the nineteenth-century culture of urban representation was inextricably entangled with the logic of colonialism.[29] The more the city presented itself as artificial representation, the more powerful became a picture of a natural world existing outside this space of representation. The construction of what he calls the 'world-as-exhibition' had the effect of dividing the world in two: 'on the one hand a material dimension of things themselves, and on the other a seemingly separate dimension of their order or meaning'.[30] He highlights the colonial implications of this binary: the material world of things themselves, lacking the meaning and order supplied by the exhibition, was an essentialized and exotic Orient. The vast volume of travel writing produced in the nineteenth century served to make that chaos legible, turning the Orient itself in a vast exhibition. Derek Gregory cites travellers' descriptions of Egypt as 'an open-air museum where temples and tombs are arranged like shop windows for public inspection'.[31] Travel writing emphasized the wild and untamed nature of the Orient at the same time as heroically exerting its visual mastery over its domain and creating a legible order out of it. Such travel writing reduced colonized space to a vast open-air museum, and colonized people became exhibits in an enormous exhibition of the real.

The most extraordinary examples of this reduction of living peoples to representational exhibits (and clear exemplars of the intersecting logics of museum display and imperial display), were the Expositions Universelles. These vast spectacles presented to the public supposedly complete, encyclopaedic displays of art, science, history and anthropology.[32] Like the museum, the exhibitions offered universal surveys of comprehensive, authoritative knowledge. Crowds were to be educated, not by selective instruction, but through exposure to a complete representation of a certain area of knowledge in one spectacular exhibit.

---

29    Timothy Mitchell, *Colonising Egypt* (Berkeley, 1988), Timothy Mitchell, 'Orientalism and the Exhibitionary Order', in Nicholas Dirks (ed.), *Colonialism and Culture* (Ann Arbor, 1992).

30    Mitchell, 'Orientalism and the Exhibitionary Order', p. 302.

31    Derek Gregory, 'Scripting Egypt: Orientalism and the Cultures of Travel', in James Duncan and Derek Gregory (eds), *Writes of Passage: Reading Travel Writing* (London and New York, 1999), p. 134.

32    Paul Greenhalgh, *Ephemeral Vistas: The Expositions Universelles, Great Exhibitions and World's Fairs, 1851–1939* (Manchester, 1988).

One example of this was Charles Garnier's popular 'History of Human Habitation' exhibit at the 1889 Exposition, which displayed a street of 39 houses, each constructed in authentic manner, and each representing a specific culture and a stage in world architecture, from prehistoric times onwards. By wandering along a single street, visitors could take in at glance the entire history of human architecture, presented in linear, evolutionary terms. A similar insight was supposed to be drawn from viewing entire villages of peoples from colonized territories who had been brought to Paris in order to make visible the path of human evolution. By exhibiting these peoples, evolutionary temporalities were presented to urban dwellers, and the place of the French population on this evolutionary scale of natural life made visible.

The repetition of the world-as-exhibition in travel writing supported techniques which made the historical life of the nation visible in the passage between raw, chaotic nature and the orderly representations of the modern metropolis. The travel writing of the *Chat Noir* turned this masterful gaze back upon the city. However, it did not simply reverse the binary opposition between life and representation, adopting a vitalistic celebration of unadulterated nature in opposition to the modern culture of artificial representation – a stance would be taken up in later years by avant-garde movements such as Surrealism.[33] Nor did its landscapes merely portray Paris as an inchoate and primitive space of uncivilized nature. Rather, by drawing once again on an aesthetic of failure and error, continually deviating from the smooth passage between sensation and representation, they troubled the epistemological framework upon which these distinctions were built.

This can be made clearer by considering the Chat Noir's Parisian landscapes in terms of the ways in which they modified the posture of the flâneur. The figure of the flâneur had first emerged in the early to mid-nineteenth century as an archetypal dweller and observer of urban space, a heroic individual with the leisure to wander, watch and browse and supposedly possessing a heightened ability to decipher the dense web of the urban text.[34] The flâneur embodied the epistemic framework of the world-as-exhibition, with his ability to exercise a visual mastery of space, and his ability to transfigure the city into a museum-like, wholly legible space of exhibition and display, a representational space to be read and decoded. Mary Gluck argues that, following Haussmann's modernization of Paris, and the city's transformation into a rationalized, predictable, easily legible urban fabric, the role of the flâneur changed, transforming from a 'popular' flâneur to the kind of 'avant-garde' flâneur celebrated by Baudelaire.[35] The new

---

33   Michael Sheringham, *Everyday Life: Theories and Practices from Surrealism to the Present* (Oxford, 2006), pp. 66–70.

34   Susan Buck-Morss, 'The Flâneur, the Sandwichman and the Whore: The Politics of Loitering', *New German Critique* 39 (1986), Keith Tester (ed.), *The Flâneur* (London, 1994), Elizabeth Wilson, 'The Invisible Flaneur', *New Left Review* 191 (1992).

35   Mary Gluck, 'The Flâneur and the Aesthetic: Appropriation of Urban Culture in Mid-19th-Century Paris', *Theory, Culture & Society* 20(5) (2003), Mary Gluck, *Popular*

kind of 'avant-garde' flâneur no longer used his imagination in order to make the urban experience legible; instead, he defended the imagination *against* the city's over-determined semiotic web. He no longer created representations out of a chaos of sensation. Rather, his skill was to harvest the dynamic sensations embedded beneath the dead weight of the urban text. The history of flânerie in the nineteenth century, then, replicates the binary structure of the world-as-exhibition. The 'popular' flâneur created representations out of chaotic sensations; the 'avant-garde' flâneur extracted sensations from the city's iconographic layers of meaning.

The posture of the *Chat Noir*'s travel writing, however, was different to each of these stances. By adopting the posture of the colonist, the explorers portray the city of Paris as a chaos of sensation, needing to be made intelligible by the heroic explorer. They take up an attitude, therefore, that resembles that of the 'popular' flâneur. The humorous element of the writing, however, results from the *failure* of the explorers to make the city legible. They undermine the city's claim to possess a legible layer of meaning testifying to its modernity, authority and power. In this respect, the explorers are akin to the 'avant-garde' flâneur, taking apart the representational fabric of the city in order to encounter the more vital layers of experience bubbling underneath. Yet the explorers fail in this task as well: rather than encountering a life that is more primitive and vital, they encounter only death and sterility. They find themselves immersed in a space that is dead, alien, and hostile to the capricious naïveté of the native Montmartrois. Their roving gaze reduces the city's most important monuments to spaces of death. At one point, for example, the explorer encounters the symbolic heart of Paris, the Cathédral Notre-Dame, which occupied a site close to the Morgue – itself a popular fin-de-siècle spectacle – and conflates the two.[36]

> One morning, I was gazing at a strange and low monument, built at the bifurcation of two branches of the Seine, just behind a proud and elevated monument, decorated by a large belfry and two immense towers with high windows. I addressed myself to an old woman … and asked her in Parisian – this dialect has become familiar to me – 'What is this building?' (pointing to the building on the right), 'and what is that construction?' (indicating the one on the left) … The aged person replied: 'This is Notre-Dame, and that is the Morgue. We put corpses in this one, and say their mass in that one'. I thanked the old woman, and left her, wondering why these two buildings possessed, the one a female name, and the other, a complete absence of amenity for its fellow citizens. In Paris, cadavers have the morgue, proof of the complete barbarity of Sequanaise populations; and in addition they have Notre-Dame. I believe that, in

*Bohemia: Modernism and Urban Culture in Nineteenth-Century Paris* (Cambridge, MA, 2005).

36   On the morgue as spectacle, see Vanessa Schwartz, *Spectacular Realities: Early Mass Culture in Fin-De-Siècle Paris* (Berkeley, 1998).

order to save saliva, the inhabitants of Paris should say, in speaking of these two buildings, built on the same ground: 'Our Lady of the Morgue'.[37]

The explorers fail at either form of flânerie. They are incapable either of extracting sensations from representation, or representations from sensation, and the humour of their exploits results precisely from this interruption of the passage between the two. The effect of this is to problematize the passage itself, and the epistemological framework of the world-as-exhibition that it presupposes. The construction of the world-as-exhibition relied upon a contrast between life and representation, nature and culture, which naturalized the passage between the one and the other and made invisible the power relationships involved in this passage.

Imperial travel narratives portrayed the landscapes they encountered as wholly natural (and thus apolitical), but also as wild, chaotic, and only graspable through conversion into coherent representations. Here natural life was assumed to be something that required organization through representation in order to be made coherent and graspable – just as Kant had theorized nature in the *Critique of Judgement*.[38] Here the power of representation appeared as something necessary, thus enabling the relations of authority implicit in it to remain invisible. In both cases, techniques of making visible were at the same time techniques of making invisible. It was this process of making invisible that the landscapes of the *Chat Noir*, in turn, made visible once again. Through its humorous performance of error – both the failure to create a legible space of museum display, and also the failure of its explorers to decode the city and demonstrate the distance between nature and civilization – the *Chat Noir* ridiculed bourgeois claims concerning the possibility of organizing life through representational means. Through its constant and deliberate lapse into error and misunderstanding, it evoked an experience of a truthful life *outside* represented life, a vitality generated from wandering or erring.

---

37　Jacques Lehardy, 'Lettres D'un Explorateur', *Le Chat Noir*, 18 March 1882.

'Un matin, je contemplais un monument étrange et bas, construit à la bifurcation de deux branches de la Seine, juste derrière un monument fier et élevé, orné d'un grand clocher de deux immenses tours percées de hautes fenêtres. Je m'adressai à une vieille femme ... et lui demandai en parisien, – ce dialecte m'est devenu familier: – Quelle est cette baraque? en désignant le bâtiment de droite, et comment nommez-vous cet édifice? en indiquant la maison de gauche. La personne âgée me répondit: Ca, c'est Notre-Dame, et ca, c'est la Morgue. Ici, on met les macchabées, et là, on leur dit la messe. – Je remerciai la vieille dame, et je la quittai en me demandant pourquoi ces deux constructions portaient, l'une un nom féminin, et l'autre celui d'un défaut constitué par une absence complète d'aménité envers ses concitoyens. A Paris, les cadavres ont de la morgue, preuve de la barbarie complète des populations séquanaises; puis ils ont Notre-Dame. Je crois que, pour économiser la salive dans leur locutions, les habitants de Paris pourraient dire, en parlant de ces bâtisses, construites sur la même terrain: Notre Dame de la Morgue'.

38　Immanuel Kant, *The Critique of Judgment*, trans. J. H. Bernard (New York, 2000).

**Life, Error and Lived Experience**

To be understood adequately, the *Chat Noir*'s emphasis on error and a 'poetics of failure' must be interpreted in the context of dominant theories of experience in the late nineteenth century – and in particular, the tradition of thought that deployed a critique of alienation.[39] The concept of alienation – associated most stronglywith Marx – derives from the Hegelian theory of experience (*Erfahrung*).[40] For Hegel, the highest good is what he calls 'unity of life' (*Einheit des Lebens*).[41] The highest good consists in achieving unity, wholeness or harmony in all aspects of our being – with oneself, with others, and with nature. The biggest threat to this unity is division or alienation, where the self finds itself divided from itself, from others, and from nature. Its goal is to overcome these divisions and achieve unity, so that it is once again 'at home in the world' (*in die Welt zu Hause*) and can achieve full self-knowledge. Since the unity of Spirit is based upon the structure of organic life, the theory of alienation is conceptually tied to a critique of mankind's fall from life. Alienation is a form of separation – between self and other, between subject and object, and between knowledge and experience. When experience is no longer a manifestation of life, but is separated from it, then mankind becomes alienated. When a distance emerges between life and experience, it is no longer possible to be 'at home in the world'.

Whether being 'at home in the world' in this way (that is, participating in the organic immanence of life or the Absolute) is either desirable or possible, however, is open to debate, and the Chat Noir's emphasis on the creativity of 'erring' points towards an alternative ethos towards the life of the city. Here, the aim was not to eliminate the separation between life and experience but to deliberately explore forms of alienated life, with the aim of discovering new forms of more vital and energetic life. One way of reframing this issue, then, is to reconceptualize life, not in terms of a stable equilibrium between a living body and its environment (a way of being 'at home' in the city), but, as itself a form of error.[42] Here we can draw on the work of Georges Canguilhem, who in response to the arguments of ethologists

---

39    On the poetics of failure, see Sara Jane Bailes, *Performance Theatre and the Poetics of Failure: Forced Entertainment, Goat Island, Elevator Repair Service* (London, 2011), Judith Halberstam, *The Queer Art of Failure* (Durham, NC, 2011), Lisa Le Feuvre (ed.), *Failure: Documents of Contemporary Art* (Cambridge, MA, 2010), Róisín O'Gorman and Margaret Werry, 'On Failure (on Pedagogy): Editorial Introduction', *Performance Research* 17(1) (2012).

40    For the classic Marxist account of alienation, see Karl Marx, *Economic and Philosophical Manuscripts of 1844* (Moscow, 1961).

41    Georg Wilhelm Friedrich Hegel, *Phenomenology of Spirit*, trans. A. V. Miller (Oxford, 1977). See Frederick Beiser, *Hegel* (London and New York, 2005), p. 37.

42    See Georges Canguilhem, *On the Normal and the Pathological*, trans. Carolyn Fawcett (Dordrecht, Holland, 1978), Michel Foucault, 'Life: Experience and Science', in Paul Rabinow (ed.), *Aesthetics, Method, and Epistemology: Essential Works of Foucault 1954–1984, Volume Two* (New York, 2000).

or molecular biologists that life is tightly programmed by milieu and genetic code, focuse on living beings' errors and victories over error. Many such errors arise from a difficulty with adapting to a milieu.[43] According to Canguilhem, 'Mankind makes mistakes when it places itself in the wrong place, in the wrong relationship with its environment, in the wrong place to receive the information needed to survive, to act, to flourish'.[44] To survive, it *errs* – it moves and adapts. And this is perhaps a better characterization of the kind of life mobilized through the humourous performances and literature of Montmartre: not an equilibrium between organism and milieu (the creation of a 'home in the world'), but organisms' *errors*, their creative responses to finding themselves out of place, in an environment that is indifferent to them. When life is theorized like this, alienation – insofar as it is defined in the Hegelian manner as the separation of experience from the unity of life – becomes less a fall *from* life than the very condition of possibility of life. This is precisely the point of those Deleuzian characterizations of life, for example, which argue that, rather than inhering in organisms, life is 'everything that passes *between* organisms'.[45] Life is a state of dis-equilibrium which forces a change in the relation between an organism and its milieu, such that both organism and milieu are forced to (partially) decompose and recompose themselves.

Interpreted through this analytical framework, we can recognize that an important aim of the Chat Noir's forms of humorous counter-display was not just to criticize the ways in which life was represented in cultural forms such as museum display and travel writing, but also to experiment with developing new experiences of life – forms of *erring* life, where the experience of alienation characteristic of the modern city was recast as the condition of possibility for a new experience of creative, unrepresentable, and directly encountered life.

## Conclusion

The *Chat Noir Guide* and the *Voyages de Découvertes* imagined the Chat Noir in terms of a very specific approach to the nineteenth century culture of display. From these texts it is possible to extract an ethos towards representational life best characterized as a form of 'counter-display'. The Chat Noir portrayed Montmartre as a theatre in which a dispersal of the modern ordering of time (exemplified by the museum) and nature (exemplified by travel writing) could be staged. In doing so, new forms of direct, modern, dis-organized experience could be engineered. This ethos towards urban life was oriented towards disrupting the techniques by

---

43   See Georges Canguilhem, 'The Living and Its Milieu', *Grey Room* 3 (2001).
44   Paul Rabinow, 'Introduction: A Vital Rationalist', in François Delaporte (ed.), *A Vital Rationalist: Selected Writings from Georges Canguilhem* (New York, 1994), p. 20.
45   Gilles Deleuze and Félix Guattari, *A Thousand Plateaus: Capitalism and Schizophrenia, Volume Two*, trans. Brian Massumi, 2nd ed. (London and New York, 2004), p. 550.

which life was organized and ordered into a functional, stable totality. Rather than representing life as something that could be organized, through representation, in a rational, ordered evolutionary sequence, this was an attitude that was determined to expose the arrogance and hubris of a faith in positive knowledge and scientific and industrial progress.

However, this ethos was not simply oriented towards deploying the authority of a purely non-representational, dynamic space of authentic lived experience. Through its parodies of colonial exploration, the literature of the Chat Noir revealed the emptiness, sterility and attenuation of experience within the city's most 'modern' monuments and boulevards. Not only did modern urban representation fail, but so did modern experience. Paris, as a representational space, was exposed as an empty shell, a life with no life – an experience of nothingness. Above all, the Chat Noir's use of counter-display was oriented towards exploring a different experience of life, not as something stable and ordered, but as a form of error, an interruption of the given ordering of visibilities and representations. Through humorous counter-display, a novel form of urban vitality might perhaps emerge, a vitality that existed *outside* life as it was specified within the power relations of modern urban display. The Chat Noir's forms of counter-display refused the Kantian model by which life is only perceptible when organized through representation, replacing it with a model where life could be experienced as precisely what tested the limits of established forms of representation.

This stance of counter-display encompassed a rejection of the dominant attitude that saw representation as the most appropriate means by which to engage with and organize the life of the city. It was opposed, for example, to the stance of the museum curator or the heroic traveller. Rather than merely reverting from representation to the non-representational – and thereby confirming the epistemic divide between reality and representation – these Chat Noir texts exposed the failures of dominant modes of visibility. They exposed modernity's dominant mode of presentation itself, and, in doing so, made apparent the existence of hidden relations of power and authority within it. The Chat Noir's forms of representation focused heavily upon the failure of representation to achieve what it claimed it could – to create total orderings of life, time and history – and, in doing so, revealed the cracks and fissures within this mode of experience. By bringing the act of representation itself to the foreground, they made clear that no direct, unmediated bridge between representation and reality was possible. They presented the act of presentation itself, and in doing so, revealed it to be, not just an act of making visible, but also an act of making invisible, of hiding the relations of power and authority involved in the production of knowledge and the representation of history and nature. This counter-representational stance towards the life of the city was a stance that aimed to create the conditions for making for new forms of life and vitality discernible. This raised the question, however, of perception: how could new forms of life beyond organic life and the conventional structures of sensation be made visible? From what ghostly sources would these forms of life spring?

# Chapter 7

# The Silent Horror of the Night:
# Supernatural Landscapes
# and the Life of the Senses

Two elements of the dominant attitudes towards the life of the city in late nineteenth century Montmartre – the *affective* life of the city and the *representational* life of the city – have now come into focus. Beyond these emphases upon affirmation and counter-display, however, Montmartre culture also focused heavily on the *perceptual* life of the city. Indeed, it was these experiments that gave cabarets such as the Chat Noir their greatest successes. In this chapter I am interested in the relationship between humorous performance, the life of the city, and the role of sensory experience in evoking forms of perceptual life beyond the limits of biological life. Investigating ghostly, supernatural and 'synaesthetic' literary and performed landscapes, the chapter will explore how these landscapes evoked and parodied the limits of sensory life. Its experiments with the distribution of the senses, we will see, tested biopolitical knowledges concerning the nature of healthy bodily perception and its relation to the urban environment, and were a challenge to the terms by which urban life, and the forms of sensation appropriate to it, could be experienced. The literary and performed landscapes of the Chat Noir, as we will see, oscillated between two different forms of extra-sensory experience: an experience of nothingness, of an incorporeal and ghostly urban milieu; and an experience of totality, a synaesthetic experience of a sensory unity beyond the natural body. Through these forms of experience, the Chat Noir established a novel stance towards the 'unworking' of the sensory body.

## The Anaesthetized City

In contrast to the usual emphasis in the culture of the Third Republic on documenting the experiential life and organic vitality of the modern city, many Montmartre writers constructed disturbing poetic landscapes of dead, sterile, anaesthetized urban space. These literary landscapes explored an experience of death, of nothingness, questioning the understanding of life that was mobilized in the Republic's ongoing urbanization projects. A poem by Edmond Haraucourt called 'The Dead City' (La Cité Morte), for example, is characteristic of this attitude:

Under a tranquil, flatly blue sky
Where the air sleeps, without warmth or vital force,

> A deserted city with pale walls spreads out,
> Sad as Death and vast as God.
>           …
>
> No grass; no flowers; no green-leafed trees:
> Everything is petrified in a giant slumber,
> And one can almost see black nothingness gaping
> In the deep frames of open windows.
> Everything is shut up; in a flash, life has left
> This fantastic world, where crowds once made a din;
> The large thresholds wait for throngs of people,
> But those who pass over them are dwellers in eternity.[1]

The poem expresses a way of seeing urban space that is far removed from that of the dazzled, roving eye of the flâneur or the modern consumer. Its gaze strips the city of every thread of life. And if at first the reader supposes that this ghostly city of the dead exists as a purely imaginary urban landscape, one which bears no relation to that of the real city, then this supposition is rejected in the final two verses. Here the narrator reveals this anaesthetized space to be an emotional landscape of modern urban subjectivity. The city of the dead becomes entangled with the Paris of modernity:

> Oh my heart! Empty, lifeless and desolate city,
> You will sleep endlessly, for as long as I live:
> My boredom will oversee your troubled slumber,
> Like a plaintive marble at the side of a mausoleum.
> Sleep in the calming forgetfulness of once-loved dreams:
> Let us await the death which brings new blood to things,
> Since all our hopes, opening their rosy wings,
> Have departed forever in fearful flight![2]

---

1    Edmond Haraucourt, 'La Cité Morte', *Le Chat Noir*, 27 May 1882.
     'Sous la tranquillité d'un ciel platement bleu
     Où l'air dort, sans chaleur et sans force vitale,
     Une ville déserte, aux murs pâlis, s'étale,
     Triste comme la Mort et grande comme Dieu.
              …
     Point d'herbes; point de fleurs; point d'arbre aux feuilles vertes.
     Tout s'est pétrifié dans un sommeil géant,
     Et l'on croit voir bailler les noirceurs du néant
     Dans le cadre profond des fenêtres ouvertes.
     Rien n'est clos: c'est d'un coup que la vie a quitté
     Ce monde fantastique où bruissait la foule;
     Les grands seuils inusés attendent qu'on les foule,
     Mais ceux qui passaient là sont dans l'éternité'.
2    'O mon cœur! cité vide, inerte et désolée,
     Tu vas dormir sans trêve et tant que je vivrai:

This deeply pessimistic landscape of a phantasmic, lifeless city evidently departs from the lively, vital, energetic kinds of city imagined in orthodox republican discourse. Modern urban life is transfigured here into a space of death and sterility, a landscape in which genuine experience is impossible. Life can only emerge from 'the death which brings new blood'.

This pessimism towards experiential life is echoed in a poem by the Montmartre poet Maurice Rollinat, who was well-known at the time for his macabre, ghostly, nightmarish poetry. In a poem titled 'Néant' (Nothingness), life itself becomes a drain on the vitality of the self:

> Life, rapacious entomber,
> Inhales you like a siphon;
> It empties you to the greatest depth,
> And leaves you a shell
> Blown around by Time in space ...[3]

Space, here, does not have any positive qualities whatsoever, but is reduced to a metaphysical abstraction, drained of genuine experience. Life itself is abstract, violent and destructive.

One effect of these ghostly poems, with their observation of the forms of death that are immanent to natural and everyday urban life, is to question the dominant emphasis on the natural vitality of the city. In these emotional landscapes, the only circulation of life is a deathly life, something static, sterile, and anaesthetized. The life of the modern city, they imply, is not genuine life, but only a ghostly abstraction of it. This critique of late nineteenth-century urban experience enabled the artists of Montmartre to lay claim to be offering a more fully vital, *living* experience in their cafes and cabarets. In order to do this, they started experimenting with challenging the established boundaries between poetry, music and performance, and in particular with placing greater emphasis on the body and embodied performances.

Montmartre's cabarets were highly eclectic, forming a venue for monologues, music, theatre, graphic arts, performed poetry and many other cultural forms.

---

Mon ennui veillera sur ton sommeil navré,
Comme un marbre plaintif au bord d'un mausolée.
Dors dans l'oubli calmant des rêves que j'aimais:
Nous attendrons la mort qui rajeunit les choses,
Puisque tous nos espoirs, ouvrant leurs ailes roses,
Dans leur vol effrayé sont partis pour jamais!'

3  Maurice Rollinat, 'Le Néant', *L'Abîme* (Paris, 1886).
'La vie, enterreuse rapace,
Vous aspire comme un siphon;
Elle vous vide au plus profond,
Et vous laisse une carapace
Que le Temps souffle dans l'espace ...'

Rollinat in particular was well known for the idiosyncratic style of his performances. Accompanying himself on the piano, reciting or singing his poetry in otherworldly voices, Rollinat *performed* his poetry, striving to conjure spirits and phantoms from the depths of the night. As one observer noted, 'Rollinat's originality comes from his being neither a poet, nor a musician, nor a speaker of verse, but all three at once. To judge it, you need to hear the three Rollinats united as one. Just as his music loses its flavour when played or sung by another, so the verses in the music are no longer there'.[4] The effectiveness of Rollinat's poetry was tied to the intense lived experiences that he could engineer through performing it in a specific time and place. Barbey d'Aurevilly wrote that, 'When [Rollinat] speaks and sings his verses, with that strident voice which seems no longer to emanate from human entrails ... He is positively the embodiment of the devil ... He is a young man with a slender elegance, fine, pure, handsome features, and a pallor that is more distinguished than sepulchral. But all of this burns and transforms when he is seized by these three hands of poetry, music and gesture, and we no longer recognize him'.[5]

The cabaret form encouraged work that enabled different art forms to collide and coalesce. Poetry could no longer exist as mere words on a page but had to be performed in order to capture its most affective, non-representational qualities. Word had to be brought to presence through gesture, speech and song. In Montmartre, observes Elodie Gaden,

> *poïesis* – the manufacture of the poem – thus does not stop in the poet's study but is completed on the stage, where the verses are voiced and embodied ... In the cabarets, the poet utters and *incarnates* – with gestures and mimicry – his poems, rather than 'declaiming' them or 'reciting' them, as Mallarmé still did. The Hydropathes, in the Chat Noir and elsewhere, experimented with other modes of oral, indeed corporeal, realizations of the poem – improvization included – leading to an entirely new conception and practice of *poetry as a performing art.*[6]

---

4   Leo Trézinek, in *Lutèce*, 8–15 May 1885.

5   Cited in Jean-Jacques Lévêque, *Les Années Impressionnistes* (Courbevoie, 1990), p. 98.

'Quand il dit ses vers et qu'il les chante, avec cette voix stridents qui semble ne plus sortir d'entrailles humaines ... il a positivement le diable au corps ... c'est un jeune homme de gracile élégance, de pâleur plus distinguée que sépulcrale, aux traits fins, beaux et purs. Mais tout cela flambe et se transfigure quand il est saisi par ces trois mains de la poésie, de la musique et de la mimique et on ne le reconnaît plus'.

6   Élodie Gaden, 'Mutations Des Pratiques Poétiques: 1878–1914, Maurice Rollinat, Marie Krysinska, Valentine De Saint-Point' (Mémoire de Master 2 de recherche, Université Stendhal, 2009), p. 9. 'Le poème, non content d'avoir été "créé" (écrit, tracé), doit encore être "interprété" (proféré, vécu): la *poïesis* – la fabrication du poème – ne s'arrête donc pas dans le cabinet de travail du poète mais trouve son achèvement sur la scène, où les vers sont mis en voix, et en corps, d'une manière très différente de celle des salons et banquets d'alors.

The cabaret form, then, was established as a way of creating new forms of intense, embodied, dynamic sensation – a possibility lacking in more hegemonic, anaesthetized spaces of the city. One way in which they attempted to offer a counter to the sterility and banality of modern urban life was to create performances spaces in which art forms could interact in new ways, creating novel registers of sensation and lived experience.

These ideas raise further questions, however. If the supposed 'life' of the modern city, rigidly controlled and organized, is really a kind of death, a form of nothingness in which genuine experience is impossible, then might the mysterious abyss of death itself not offer the best route towards stylizing new forms of urban vitality? Are there new forms of life to be found *beyond* life, beyond the limits of natural perception?

## Spectral Life

The diagnosis of modern life as a form of death provided a backdrop to artistic experiments with apprehending supernatural and non-rational forces that circulated, invisible but still perceptible, within the deepest layers of bodily experience. Maurice Rollinat became a figurehead for this form of poetic experimentation, which in some ways anticipated the experiments of the Surrealists in later decades. Rollinat evoked a world of elemental, irrational, terrifying forces beyond the realm of organic perception. He was fascinated by forms of life that pulsed *beyond* the limits of natural life, but which were faintly discernible within the most intense forms of sensation, as well as within the darkest recesses of sleep.

Rollinat's poetry, evoking themes in Baudelaire and Poe, went further than either in his exploration of death, putrefaction and diabolism. In one poem, performed by Rollinat at the piano, we encounter another kind of music:

> Entirely naked, she was seated at the harpsichord;
> And while savage winds howled outside
> And Midnight sounded like a vague alarm bell,
> Her cadaverous fingers flew around the keys.
>
> A pale night light illuminated sadly
> The room where this tragic scene was occurring,
> And sometimes I heard a low groan
> Mingling with the chords of the magic instrument

---

Dans les cabarets, le poète profère – jusqu'au cri – et *incarne* – avec gestes et mimiques – ses poèmes, plus qu'il ne les "déclame" ou les "récite", comme le disait encore Mallarmé. Aux Hydropathes comme au Chat Noir et ailleurs, s'expérimentèrent d'autres modes de réalisation orale, voire corporelle – improvisation comprise –, du poème, menant à une conception et à des pratiques entièrement nouvelles de *la poésie comme art de la scène'*.

Oh! Magic indeed! Because it seemed to speak
With the thousand voices of an immense harmony,
So broad that it must flow
From a sea that is musical and full of genius.

My spectral lover, reached by death,
Thus played in front of me, livid and violet,
Her hair, so long, and blacker than remorse,
Falling back softly on her living skeleton.[7]

Beauty and genius, here, come from beyond the grave, from a living death. The artist must reach beyond the natural world in order to grasp hold of the truth. But it is also in this world that man's darkest fears are lurking. Another poem, titled 'Peur' (Fear), is characteristic in its fascination with spectral worlds, where shadowy forms dissolve and reshape themselves in extraordinary new ways. The poem starts with the fall of night and the fall into troubled, nightmarish sleep. It is narrated by a presence that is both 'invisible' and 'ubiquitous', ordering and disordering the world of dreams.

As soon as the sky is veiled
And the evening, brown weaver,
Puts itself to creating its cloth
In the recurring mystery,

I subject man to my whim,
And, queen of ubiquity,
I make him convulse and his hair prickle,
Through my invisibility.

If sleep shuts his eyelids,
It commands a sick nightmare

---

7    Maurice Rollinat, 'L'amante Macabre', *Les Névroses* (Paris, 1885).
     'Elle était toute nue assise au clavecin;
     Et tandis qu'au dehors hurlaient les vents farouches
     Et que Minuit sonnait comme un vague tocsin,
     Ses doigts cadavéreux voltigeaient sur les touches.
     Une pâle veilleuse éclairait tristement
     La chambre où se passait cette scène tragique,
     Et parfois j'entendais un sourd gémissement
     Se mêler aux accords de l'instrument magique.
     Oh! magique en effet! Car il semblait parler
     Avec les mille voix d'une immense harmonie,
     Si large qu'on eut dit qu'elle devait couler
     D'une mer musicale et pleine de génie.
     Ma spectrale adorée, atteinte par la mort,
     Jouait donc devant moi, livide et violette,
     Et ses cheveux si longs, plus noirs que le remord,
     Retombaient mollement sur son vivant squelette'.

To press down on his breast
Like a clamping plate on a stone.

I go into his cold corridor,
I carry myself to his landing
And suddenly, as if with a finger,
I knock knock knock at his door.

On his table, beside an owl,
Is sitting a severed head
With an insane grin
And the gaze of a doll.

He sees approaching, with crawling steps,
A lady of deathly pallor,
With snakes for hair
And a shroud for a dress.

Then I extinguish his lamp, and sit
At the edge of his bed, which digs itself
A cadaverous form
Which tickles his two feet.[8]

In the dream-filled realm of the night, the world beyond vision, beyond the sensible, starts to make itself faintly discernible. Forces reveal themselves that overturn space, time and measure:

---

8 Maurice Rollinat, 'La Peur', *Les Névroses* (Paris, 1885).
   'Aussitôt que le ciel se voile
   Et que le soir, brun tisserand,
   Se met à machiner sa toile
   Dans le mystère qui reprend,
   Je soumets l'homme à mon caprice,
   Et, reine de l'ubiquité,
   Je le convulse et le hérisse
   Par mon invisibilité.
   Si le sommeil clôt sa paupière,
   L'ordonne au cauchemar malsain
   D'aller s'accroupit sur son sein
   Comme un crapaud sur une pierre.
   Sur la table, ainsi qu'un hibou,
   Se perche une tête coupée
   Ayant le sourire du fou
   Et le regard de la poupée;
   Il voit venir à pas rampants
   Une dame au teint mortuaire,
   Dont les cheveux sont des serpents
   Et dont la robe est un suaire.
   Puis j'éteins sa lampe, et j'assieds
   Au bord de son lit qui se creuse
   Une forme cadavéreuse
   Qui lui chatouille les deux pieds'.

I distort all noises,
I corrupt all forms,
And I transform the most enormous mountains into deep shafts.

I scramble time and place;
Under my fantastic will,
Summit becomes middle,
And measure is elastic.

I bring torrents to a standstill,
I harden water, I melt marble,
And I uproot trees to make
Wandering Jews of them.[9]

Here Rollinat gives full expression to a ghostly confusion of spatiality and temporality, a world in which nothing can be held still, every form dissolving into another. The atmospheres that Rollinat created provoked pleasurable shivers of terror in his audiences, and emphasized Montmartre's role as a space in which wholly new kinds of experience could be encountered. Montmartre was to become a uniquely *phantasmagorical* space of urban modernity.

### Shadows, Silhouettes, and the Confusion of the Senses

During its early years, the Chat Noir remained a very small institution with a marginal place in the cultural topography of Paris. Although it soon became a fashionable space for bohemian artists to gather, it was only with the invention of its shadow theatre, or 'ombres chinoises' (Chinese shadows), that the Chat Noir started to become a widely popular, commercially-successful venue. These shadow plays proved so successful that they would eventually go on tour around the country, forming an important element of fin-de-siècle popular culture.[10] These shadow plays, like the ghostly landscapes discussed above, evoked an urban landscape beyond the senses. This time, however, they gestured, not towards the nothingness of the 'dead' city, but towards a new form of sensory totality, a totality beyond the natural unity of the sensing body.

---

9   Ibid.
    'Je dénature tous les bruits,
    Je déprave toutes les formes,
    Et je métamorphose en puits
    Les montagnes les plus énormes.
    Je brouille le temps et le lieu;
    Sous ma volonté fantastique
    Le sommet devient le milieu,
    Et la mesure est élastique.
    J'immobilise les torrents,
    Je durcis l'eau, je fonds les marbres,
    Et je déracine les arbres
    Pour en faire des Juifs errants ;'

10  P. Jeanne, *Les Théâtres D'Ombres a Montmartre de 1887 a 1923* (Paris, 1937).

Figure 7.1    Henri Rivière, image from the shadow play 'La Tentation de Saint Antoine', as reproduced in Henri Rivière, *La Tentation de Saint Antoine, Féérie à Grand Spectacle, en 2 Actes et 40 Tableaux* (Paris: Chat Noir, 1887), p. 11

Figure 7.2    Henri Rivière, image from the shadow play 'La Marche d'Etoile', as reproduced in Georges Fragerolle, *La Marche à L'Étoile, Mystère en 10 Tableaux* (Paris: Chat Noir, 1890), p. 29

The technique of shadow plays was invented by a Chat Noir regular called Henri Rivière.[11] It involved placing silhouettes of buildings, crowds and other elements of an urban landscape within a wooden frame at several distances from a screen. The closest created a completely black shadow, and the next ones created successive gradations of grey, thereby conveying a sense of recession into space. The silhouettes, made first from cardboard and then from zinc, were moved across the screen on runners. Behind them were glass panels, also on runners, which were painted with a variety of transparent colours. Finally, behind these was an oxyhydrogen flame that served as a light source. Using large numbers of backstage assistants, Rivière conjured up highly complex, impressionistic arrangements of colour, sound and movement. The resulting shows were ghostly performances in which moving image, music and poetry mingled and coalesced. Everything was indistinct and muffled; colours, shapes and sounds mingled to create a synaesthetic atmosphere, an excess of decadent sensuality (Figure 7.1, Figure 7.2). Religious themes were often favoured, ironically drawing on Montmartre's ancient and renewed religious identity.

These multi-sensory performances contrasted strongly with the scientific, organized, hierarchical conception of healthy embodied perception that was dominant in the Third Republic. There was nothing natural or organic about this experience of perception. Instead, it gestured to a space beyond natural life. The shadow theatre seemed, to awe-struck observers, to open a window onto a new mystical realm of reality. Jules Lemaître, for example, observed that, 'the Chat Noir contributed to the "awakening of idealism". It was mystical ... the luminous wall of its puppet-theatre formed a bulls-eye opening onto the invisible'.[12] In addition, he wrote, the shadow theatre of the Chat Noir proved that 'mysticism is able to combine, very naturally, with the liveliest strength and the most Greek sensuality'.[13] Such mystical experiences, with their promise of opening sensation beyond the limits of the organic senses, were rapturously received. An account in *Le Magazine Français Illustrée* evokes the magical effect created:

> I saw the strangest nocturnal landscape that one could imagine. In the distance, mysterious shadows passed to and fro from the windows of dimly illuminated houses. Everywhere, from chimneys pitching their fantastic silhouettes and shaking their baroque weather vanes; downstairs from frightening bays of light, green and yellow; from trees suddenly appearing here and there, twisting their

---

11    See Phillip Dennis Cate, 'The Spirit of Montmartre', in Phillip Dennis Cate and Mary Shaw (eds), *The Spirit of Montmartre: Cabarets, Humor and the Avant-Garde, 1875–1905* (New Brunswick, NJ, 1996), pp. 58–9.

12    Jules Lemaître, 'Le Chat-Noir', *Les Gaîtés Du Chat Noir* (Paris, n.d.). '[L]e Chat-noir contribuait au "réveil de l'idéalisme". Il était mystique ... L'orbe lumineux de son guignol fut un œil-de-bœuf ouvert sur l'invisible'.

13    Ibid. '[Il] nous a appris que le mysticisme se pouvait allier, très naturellement, à la plus vive gaillardise et à la sensualité la plus grecque'.

black branches; from hanging gardens that seemed to have been abandoned for a thousand years; and from a great distance a vague music ... but coming from where? ... I won't try to describe it, as I think I was dreaming. Portcullises being raised and lowered, fitted with 101 electric lights; the panoramic sky unfolding as if under a fairy's wand; the blue sparks of the switches; the noise of lightning; ... the monks singing mass at the foot of the ancient cloister, and the route songs of the soldiers.[14]

What is palpable in accounts such as this is the giddy excitement of experiencing a *confusion of the senses*: an intoxicated and hedonistic sensual voyage in which the natural divisions between the senses seemed to melt away, opening onto another kind of non-natural life beyond the organic life of the body and the urban milieu. The author of this account characterizes the mélange of swirling sights and sounds as a 'symphony of colours', echoing various poems in *Le Chat Noir* with titles such as 'Symphony in Grey'.[15] This kind of sensory disorder had already been evoked in avant-garde poetry: first by Baudelaire in *Fleurs du Mal*, and then by Rimbaud, who in 1871 had written that 'the poet makes himself a *visionary* through a long, a prodigious and rational disordering of *all* the senses'.[16] Moreover, by the 1880s the controversial figure of Richard Wagner, with his ambitious dream of synthesizing the arts, unifying the sensing body, and creating a modern form of Greek tragedy, was becoming an ever more influential force in bourgeois French culture.[17]

So what was the appeal of the Chat Noir's synaesthetic theatre, and why did it prove so popular amongst the Parisian middle classes? One explanation would situate the shadow theatre within the old tradition of the phantasmagoria, which occupied a central position in the Marxist critique of mass culture and the spectacle.

---

14    Taharin, 'Au Théâtre Du Chat Noir', *Le Magazine Français Illustrée* (1891), pp. 385–6. '[J]'aperçus le plus étrange paysage nocturne qu'on puisse imaginer. Au loin, des maison faiblement illuminées derrière les fenêtres desquelles passaient et repassaient de mystérieuses ombres. Partout des cheminées dressant leurs silhouettes fantastiques et agitant leurs girouettes baroques; en bas d'inquiétantes baies de lumière, vertes et jaunes; des arbres surgissant d'ici, de là, tordant leurs rameaux noirs; des jardins suspendus qui semblaient abandonnés depuis mille ans, et très lointaine une vague musique ... mais venant d'où? ... Je n'essaierai pas de décrire cela, car je crois avoir rêvé. Les herses montant et descendant, munies de 101 lampes électriques; le ciel en panorama se déroulant comme sous la baguette d'une fée, les éclairs bleus des commutateurs, le bruit de la foudre ... Les moines chantant la messe au fond du vieux cloitre, et la chanson de route des guerriers'.

15    Marie Krysinska, 'Symphonie En Gris', in André Velter (ed.), *Les Poètes du Chat Noir* (Paris, 1996).

16    Arthur Rimbaud, *Illuminations and Other Prose Poems* (New York, 1957). On the spatialities of Rimbaud's poetry, see Kristin Ross, *The Emergence of Social Space: Rimbaud and the Paris Commune* (Minneapolis, 1988).

17    Steven Huebner, *French Opera at the Fin De Siècle: Wagnerism, Nationalism, and Style* (Oxford, 1999).

The phantasmagoria was a kind of magic lantern show in which the projectors were hidden, so that the visual effect emerged in the middle of the room as if by magic.[18] Phantasmagorias were normally populated by ghosts and spirits – entities with an uncanny presence whose origin could not be discerned. Marxist thought sees this hiding of the productive forces as a metaphor for the ideological functions of capitalist culture at large, which obscures the true relations of production of society from view. One approach to such phantasmagorical culture, then, is to dismiss it as the alienating cultural expression of capital's inner logic. Yet if we are to acknowledge that cities are as much emotional as physical spaces, it is necessary to go further than this, constructing analyses that are capable of 'taking seriously the imaginative, fantastic, emotional – the phantasmagorical – aspects of city life'.[19] And one important aspect of the urban is its ghosts and hauntings.[20] Part of the significance of urban ghosts lies in their embodiment of heterogeneous temporalities and their ability to disturb senses of place. Most importantly, ghosts demand something from the living, and take on an ability to question the limits of life itself. Ghosts possess an incorporeal presence, threatening to create a deathly absence, a hiatus within the heart of the living moment. In contrast to this impulse to see ghosts and phantasms as images of illusion and false reality, it is possible instead to approach them as very real elements of the urban environment.

Ghosts, spirits and shadows acquired a significant presence in fin-de-siècle popular and artistic culture.[21] Part of the reason for this, perhaps, is that with the ongoing conquest of nature during modernity, many people saw the supernatural world as the next domain of reality which could be captured and controlled.[22] The supernatural, in this case, was simply an extension of nature, a domain of reality that was still be conquered. Yet it is hard to explain the appeal or the aesthetics of the Chat Noir's shadow theatre in this way. Its performances seem to have had the effect, not of synthesizing the different senses into an organic, organized whole, but of *confusing* them, evoking something beyond natural life. In order to understand this, it is necessary to consider the ways in which the senses were being incorporated, during this period, into biopolitical discourses concerning the life of the city.

---

18   David Clarke and Marcus Doel, 'Engineering Space and Time: Moving Pictures and Motionless Trips', *Journal of Historical Geography* 31(1) (2005), Kevin Hetherington, *Capitalism's Eye: Cultural Spaces of the Commodity* (Abingdon, 2007), p. 62.

19   Steve Pile, *Real Cities: Modernity, Space and the Phantasmagorias of City Life* (London, 2005), p. 3.

20   Kevin Hetherington and Monica Degen (eds), *Spatial Hauntings* (Special edition of *Space and Culture*, 1(11/12), 2001).

21   N. Forgioni, '"The Shadow Only": Shadow and Silhouette in Late Nineteenth Century Paris', *The Art Bulletin* 81(3) (1999).

22   David Allen Harvey, 'Beyond Enlightenment: Occultism, Politics, and Culture in France from the Old Regime to the Fin De Siècle', *The Historian* 65(3) (2003).

## The Biopolitics of Sensation

During the final decades of the nineteenth century, new knowledges of the somatic subject were emerging that focused on the active power of the body to create experience.[23] 'What took shape in the last two decades of the nineteenth century ... were notions of perception in which the subject, as a dynamic psychophysical organism, actively constructed the world around it through a layered complex of sensory and cognitive processes, of higher and lower cerebral centres'.[24] The body came to the fore as something that played an active role in perception, synthesizing the input of its different sense organs into a perceptual whole. Perception was no longer something that enabled knowledge *of* the body; it was something that was created *by* the body.

At the same time as this body of knowledge concerning the nature of perception started to accumulate, a large number of *pathologies* of perception were discovered, in which this capacity of the body to synthesize sensory inputs became impaired, meaning that the unity of perception became liable to collapse: 'contemporary research on newly invented nervous disorders, whether hysteria, abulia, psychasthenia, or neurasthenia, all described various weakening and failures of the integrity of perception and its collapse into dissociated fragments'.[25] The ability of individuals to synthesize different units of sensory experience into a unified perceptual field was seen to be in danger. The body's ability to organize experience was in continual threat, as evidenced by the rapidly accelerating number of people diagnosed with such perceptual disorders.

This danger, however, was not just a medical matter. It was also seen as a distinctively *social* problem, because the failure of an individual body to synthesize sensations was a result of a problem with the environmental milieu in which that body dwelt. Successful synthesis of sensory experience, it was argued, was ensured when a human organism adapted successfully to its social environment.[26] The proliferation of syndromes where the unity of sensation collapsed into fragments, therefore, was directly related to the alienation of the body from its urban milieu. The fragmentation of perception was not merely a question of the health of individual bodies; it reflected on the health of the city itself. The question of the unity of sensory experience was directly and explicitly linked to the relation between the individual and the life of the city. The rise of nervous disorders was linked to the demands of the modern city upon the senses, as the city continually bombarded the body with perceptual information, leading to a sensory overload that risked overwhelming the organization of perception

---

23    Jonathan Crary, *Techniques of the Observer: On Vision and Modernity in the Nineteenth Century* (Cambridge, MA, 1990), Jonathan Crary, *Suspensions of Perception: Attention, Spectacle, and Modern Culture* (Cambridge, MA, 1999).
24    Crary, *Suspensions of Perception: Attention, Spectacle, and Modern Culture*, p. 95.
25    Ibid.
26    Ibid.

altogether.[27] The disintegration of the senses was understood as being symptomatic of the impossible demands of modern urban life.

As we have seen, an enormous amount of energy was being devoted in fin-de-siècle Paris to engineering an urban environment that would remain healthy and natural, thereby making it easy for sensing bodies to adapt to their surroundings. In a healthy environment, embodied organic perception could remain unified, ordered and organized. So the experiments of the Montmartre avant-garde can be interpreted as having been directed against precisely this way of perceiving the life of the city. In fact, the ghostly, synaesthetic, landscapes that were created and performed in Montmartre directly parodied forms of unhealthy perception that were being feverishly discussed in medical circles. These phantasmagorical landscapes can be interpreted as attempts to creatively re-appropriate the forms of urban alienation whose existence was being demonstrated by the proliferation of perceptual diseases and disorders. Phantamagorical landscapes were ways of exploring the forms of urban vitality that exceeded the orderly, clear, semantically legible forms of life that were celebrated in dominant republican discourses.

## Parody and Totality

In certain respects, the artistic experiments of the Chat Noir, with their insistence upon combining painting, poetry, music and theatre into a mystical whole, bears comparison with Wagnerian total theatre, which had a similar aim. Wagner's operas have been marked out by some critics as perfect exemplars of the spatialization of time characteristic of the modern phantasmagoria.[28] The Wagnerian *Gesamtkunstwerk* drew the arts together into a far closer weave than had been seen in Europe since ancient times. Often seen as the climax of Romanticism, Wagner's total theatre attempted to bring about a reunification of the senses, and hence to lead mankind towards a closer bond with nature than had been possible before. Rejecting modernity and embracing myth, it sought to regain an authentic relationship with nature. In this respect, Wagner wished to use art to bring humanity into a closer relationship with *totality*. Since the birth of Romanticism, the concepts of nature and totality had been closely related. Kant, for example, referred to nature as two forms of totality: the 'totality of appearances' and 'the totality of rules under which all appearances must come in order to be thought

---

27    This theme of urban analysis was to receive its most influential expression in Walter Benjamin's Freudian analysis of the role of shock effects in the destruction of experience (*Erfahrung*), as well as the urban analyses of Georg Simmel and Siegfried Kracauer. Walter Benjamin, 'On Some Motifs in Baudelaire', in Hannah Arendt (ed.), *Illuminations* (London, 1999), David Frisby, *Fragments of Modernity: Theories of Modernity in the Work of Simmel, Kracauer, and Benjamin* (Cambridge, MA, 1988).

28    For example Theodor Adorno, *In Search of Wagner* (London, 2005).

as connected in an experience'.[29] The experience of the dynamical sublime – the Romantic experience *par excellence* – was the overpowering feeling of the incapacity of sensibility and understanding to create adequate representations of nature as a totality. Wagnerian total theatre took this infinite dynamism of natural life as its ultimate model. 'Man will never be that which he can and should be, until his Life is a true mirror of Nature', he wrote.[30] This means that, 'as man becomes free when he gains the glad consciousness of his oneness with Nature; so does Art only then gain freedom when she has no more to blush for her affinity with actual Life'.[31]

I wish to suggest that the phantasmagorical performances at the Chat Noir in fact took a rather different path to that of Wagner and his followers. Once again, it is the Montmartrois emphasis on humour that helps to explain this difference. Wagner's operas were highly philosophical and somewhat humourless, and writers in the *Chat Noir* journal frequently mocked Wagner's portentous solemnity.[32] The synaesthetic shadow performances, moreover, certainly did not share Wagner's lofty ambitions. For all their sense of mystery and transcendence, they always retained a Rabelaisian, scatological humour.[33] Whereas Wagner's operas are set in mythological lands of dwarves and dragons, a popular shadow play at the Chat Noir, for example, involved an elephant walking onto a desert landscape and defecating. Another play, by Henri Somm, was set in a public lavatory, complete with stage directions for 'intestinal noises'.[34] As Elizabeth Menon recounts, the plot centred on a widow, Mme Gardetout, and her daughter, Léocadie. Mme Gardetout directs customers to their water closets, and tightly controls access to toilet paper. When love blossoms between Léocadie and a wealthy patron of the toilets, Mme Gardetout refuses to allow him to marry her daughter, on the grounds that he does not have the blessing of constipation:

> It's useless, sir. Be that as it may, my mind is made up. A constipated man, in every respect, appears to me to be the son-in-law of my dreams. Léocadie will only marry a constipated man, with whom no unwanted noises will trouble the intimacies of our house. Because, outside of our commercial relationship, we will see few people; it is among ourselves that we wish to live! What we know of society inspires in us only repugnance and contempt. I answer for Léocadie: you are, it is true, handsome, young, brilliant, but you are not constipated!

---

29    Howard Caygill, *A Kant Dictionary* (Oxford, 1995).

30    Richard Wagner, *The Art Work of the Future* (2004), p. 3.

31    Ibid., 4.

32    Georgette, 'Courrier Musical', *Le Chat Noir* 45, 46, 47, 48 (1882).

33    Elizabeth Menon, 'Images of Pleasure and Vice: Women of the Fringe', in Gabriel Weisberg (ed.), *Montmartre and the Making of Mass Culture* (New Brunswick, NJ, 2001).

34    Henri Somm, 'La Berline de L'Emigré, ou Jamais Trop Tard pour Bien Faire', *Le Chat Noir*, 1885.

Shit, here, as Menon argues, becomes a device for registering the family's contempt for the world outside, and for bourgeois life, in particular.[35] It is a highly symbolic device, registering man's corporeality and materiality. 'If we think of Symbolism as the pursuit of the Ideal, ... then shit, the obvious symbol of the material world in which we are obliged to exist, is the Symbolist substance par excellence'.[36]

The content of many of the Chat Noir plays, then, was in some degree of tension with its form. Its phantasmagorical production and ghostly, impressionistic style contrasted with a sense of humour that was as grounded, material, and as corporeal as possible. The result was not just a phantasmagoria, but a *parody* of a phantasmagoria. Rather than imitating Wagnerian aesthetics, the artists of the Chat Noir, by parodying them in synaesthetic shadow play, were attempting to transform them into something else.

In order to unpack this, we need to place the fin-de-siècle cultural interest in uniting different art-forms and exploring novel textures of perception in the context of the Romanticist philosophies that it attempted to overcome. In Romanticist thought, because art was considered to enjoy a privileged access to life, this meant that the world itself, insofar as it aspired to become more dynamic, creative and affective (that is to say, inasmuch as it turned itself towards *life*), had to become an artwork.[37] It is this totalizing idea that has been argued to have been Romanticism's most dangerous political legacy. Precisely in this spirit of Romanticism, the totalizing currents of the early- to mid-twentieth century mobilized an organic conception of political subjectivity whereby 'it is the community itself, the people or the nation, that is the [art]work following the conception acknowledged by Romanticism of the work as subject and the subject as work: the "living artwork" indeed, though this in no way prevents it from working lethally'.[38] The Wagnerian ideal of unifying the senses and making of the plural arts a singular 'art', then, is symptomatic of a drive towards totalization – towards unifying the world into a singular, living, endlessly expanding organic artwork. As Jean-Luc Nancy interprets this, the dream of Romanticism – taken up by any number of artists and politicians, from Bakunin and Marx to Mallarmé and Wagner – was of a political community based in certain respects on the model of Christian communion: humanity sharing the life of God and thereby accessing a field of pure immanence.[39]

---

35    Elizabeth Menon, 'Potty-Talk in Parisian Plays: Henry Somm's *La Berline de L'Emigré* and Alfred Jarry's *Ubu Roi*', *Art Journal* 52(3) (1993).

36    Hunter Kevil, cited in Menon, 'Potty-talk'.

37    Frederick Beiser, *The Romantic Imperative: The Concept of Early German Romanticism* (Cambridge, MA, 2003), Philippe Lacoue-Labarthe and Jean-Luc Nancy, *The Literary Absolute: The Theory of Literature in German Romanticism*, trans. Philip Barnard and Cheryl Lester (Albany, 1988).

38    Philippe Lacoue-Labarthe, *Heidegger, Art and Politics: The Fiction of the Political*, trans. Chris Turner (Oxford, 1990), p. 70.

39    Jean-Luc Nancy, *The Inoperative Community*, trans. Peter Connor, et al. (Minneapolis, 1991), p. 10.

Romanticist aestheticization involved an ethos towards creating a world in which life might be accessed as a unified, yet infinitely creative, totality.

## All and Nothing: Unworking the Work

In his campaign to become mayor of Montmartre in 1884, Rodolphe Salis, parodying Abbé Sieyès' famous 1789 revolutionary pamphlet 'What is the Third Estate?', proclaimed in an election poster: 'What is Montmartre? – Nothing. What should it be? – Everything'.[40] One critic has interpreted this in terms of an apocalyptic binary: all or nothing.[41] Perhaps, however, remembering the affirmative use of contradiction discussed earlier in relation to the experience of place (see Chapter 5), it might be better to understand it in a different way, as a conjunction: all *and* nothing. Doing so makes it possible to see better how the Chat Noir's experiments developed a specific ethos towards challenging the organization of sensory life. The aesthetics of the Chat Noir, as we have seen, drew on a strange intersection between an experience of nothingness (the spectral void at the heart of urban subjectivity) and an experience of totality (the mystical, synaesthetic experience beyond the limits of the natural body). In addition, we have seen how the experience of nothingness gestured towards another form of totality, and how the experience of totality was interrupted, through parody, and left as a form of nothingness. One thing remains unclear, however: the precise relation between the one and the other, between nothingness and totality. It is from this relation that it is possible to extract the final aspect of the Chat Noir's ethos of modernity and see how the shadow theatre attempted to challenge one form of circulation of life and replaced it with another.

Unlike Wagner, who created works of dizzying length and complexity, the cabaret culture of Montmartre produced no great works. This is not to consign it to mere historical curiosity, however. This absence of works was an important element of their approach to the life of sensation. In their landscapes of death and sensory disorder, in fact, it is possible to discern an avant-gardist impulse to take the absence of works as their very goal. If their interest in synaesthetic perception shared with Romanticism a certain drive towards totality (towards the 'all'), then their attempts to harness new forms of non-organic creativity from the abyss of 'nothingness' clearly established a distance from the Wagnerian

---

40  Rodolphe Salis, 'Poster for the Montmartre Municipal Election', (Paris, 1884). *'Qu'est-ce que Montmartre? – Rien! Que doit-il être? Tout'.* Abbé Sieyès' pamphlet asked three questions of the third estate (the common people): 'What is the Third Estate? Everything. What has it been until now in the political order? Nothing. What does it ask? To become something'.

41  Mary Shaw, 'All or Nothing? The Literature of Montmartre', in Phillip Dennis Cate and Mary Shaw (eds), *The Spirit of Montmartre: Cabarets, Humor and the Avant-Garde, 1875–1905* (New Brunswick, NJ, 1996).

phantasmagoria. This counter-Romanticist spirit would transform the notion of artistic accomplishment from the idea of completion to the idea of beginning. As a result, its experimental embodiment of synaesthetic perception would aim not to complete the natural order of the body and the immanent circulation of organic life, but to interrupt the circulation of natural life in order to stylize the experiential life of the city in new ways.

The artists of the Chat Noir did not reject Romanticism altogether. Rather, their experiments recalled a strain of it that was (at the time) neglected. This is a kind of attitude towards life, derived from the Jena Romantics, that sought to interrupt the experience of totality in order to uncover another, more vital form of life that was born of a proliferation of beginnings. This is a strain of counter-Romanticism, which, rather than affirming the infinite expansion of immanent communal life in the absolute work, instead aims at the suspension of such immanent life. 'Romanticism, it is true, ends badly', observes Maurice Blanchot, 'but this is because it is essentially what begins and what cannot but finish badly'.[42] For this reason, he argues, the most productive strain of Romanticism involves introducing a new mode of fulfilment, one which can affirm at once the absolute and the fragmentary, 'affirming totality, but in a form that ... does not realize the whole, but signifies it by suspending it, even breaking it'.[43] The goal of this subversive current of Romanticism is not the achievement of a great work, but the 'unworking' (*désoeuvrement*) of the work.[44] Unworking, here, is not the opposite of the work (that is, a process of fragmenting a complete whole). Rather, it is literally nothing: the unworked work closes and interrupts itself at the same point.[45]

In their staging of a movement between two poles, nothingness and totality, the performances of the Chat Noir experimented with a similar ethical commitment towards finding a totality *in* nothingness, rather than moving *from* nothingness *to* totality.[46] This is the 'mysticism' that Jules Lemaître referred to: the mysticism of a singularity that transcends the natural world. The performances of the Chat Noir were indeed both nothing and everything, totality and nothingness. Its performances were always 'nothing': ephemeral, transitory, unrecorded and unrepeatable. They exploited a groundless, ghostly absence that their literary landscapes made visible at the heart of the modern city. They dismantled their

---

42   Maurice Blanchot, *The Infinite Conversation*, trans. Susan Hanson (Minneapolis, 1993), p. 352.

43   Ibid., 353.

44   See Daniel Hoolsema, 'The Echo of an Impossible Future in *The Literary Absolute*', *MLN* 119(4) (2004), Ian James, 'Naming the Nothing: Nancy and Blanchot on Community', *Culture, Theory and Critique* 51(2) (2010).

45   Lacoue-Labarthe and Nancy, *The Literary Absolute: The Theory of Literature in German Romanticism*, p. 57.

46   On authority and fragmentation, see Samuel Kirwan, 'On the "Inoperative Community" and Social Authority: A Nancean Response to the Politics of Loss', *Journal of Political Power* 6(1), 2013.

own verses and paintings with wit and sarcasm. Within this nothing, however, they found a vital 'everything': an abundance of disorganized sensation, as well as a dissolution of the organized perceptual body. In this way, they sought to wrest a new form of creativity and vitality from a form of sensory life outside organic life: a vitality that was irreducible to the sterile life of the modern city.

This ethos towards 'unworking' sensation was the final aspect of the Chat Noir's attitude towards the life of the city, complementing its emphasis on affirmation and on counter-display. The performances, art and literature of the Chat Noir intervened at various levels of experience, from affective experience to representational experience to perceptual experience. In doing so, they worked upon the limits of each form of experience in order to transform its relation to the other. Representations were undermined and transformed into affects. Affects such as laughter were used in order to alter the limits of perception, making visible forms of sensation that moved beyond the natural limits of the senses. Perception was used as a way of troubling the limits of representation, making visible the possibility of new kinds of knowledge concerning the nature and limits of the life of the city. Central to all of these experiments was the attempt to create new textures of sensation that would trace, through the intensity of corporeal experience, the limits of the life of the city.

As these performances became more and more commercially successful and lost their critical edge, however, the more politically radical strands of the Montmartre community soon became disillusioned with the limitations of cultural critique. The anarchist movement, buoyed by the repatriation of convicted anarchists from the forced labour camps in New Caledonia, soon began to experiment with channelling the authority of experience through alternative, more violent routes.

# PART III
## Anarchism, Humour and Violence

## Chapter 8
# Kings of Derision:
# Anarchist Laughter and the
# Aesthetics of Violence

### Introduction

A severed head gazes down from the foot of the guillotine at the howling crowd
below. The head, frozen in rigor mortis and clotted with gore, has a terrifying
laugh frozen upon its face. It is a cold, bitter laugh, with insults and derision
on its lips. It was, until recently, attached to the body of the anarchist bomber
François Koenigstein, known as Ravachol, who had been apprehended by
police after carrying out three bombings in 1892, causing several injuries and
occasioning panic amongst the Parisian bourgeoisie. A memorable description of
his execution, published by Victor Barrucand in the anarchist journal *L'Endehors*,
elevated Ravachol to the status of a 'violent Christ'. The essay offers a revealing,
if macabre, entry point into the politics of aesthetics in modern anarchist urban
culture, which is the subject of this chapter.

Ravachol's laughing head, we are told, is to become the unlikely foundation of
a new, distinctively modern form of anarchist authority:

> Oh! This laughter before the sinister machine – a Homeric derision reverberating
> in the silence of a summer morning when all life just wanted to smile! It gives
> a funereal shudder; and it condemns for ever the social whore at the foot of the
> scaffold , attacking it with the sarcastic challenge of a criminal who has little
> care for politeness but brings into action an astonishing energy. It is as if all
> the crowd's infamy and catastrophic mediocrity were spat back in its face. The
> revolutionary's head, its jeering blasphemy unalterably frozen by the rictus of
> death, rests with a legendary authority, beautiful and purified.[1]

---

1    Victor Barrucand, 'Le Rire De Ravachol', *L'Endehors* 64 (1892). 'Oh! ce rire
devant le sinistre machine – homérique dérision répercutée dans le silence de l'estival matin
où toute vie voulait sourire! – il donne le frisson funèbre; et la catin sociale, au pied de
l'échafaud, atteinte par le sarcasme et le défi d'un criminel peu soucieux de politesse mais
qui apporte dans l'action une énergie si surprenante, est à jamais flétrie, comme si toutes ses
infamies et son irrémédiable médiocrité lui étaient jetées à la face dans ce crachat qu'elle
a mérité. Son gouailleur blasphème immuablement figé par le rictus de la mort, la tête du
révolté, belle et purifiée, demeure, avec je ne sais quelle autorité légendaire'.

The frozen laughter of Ravachol's severed head, in this grizzly description, becomes a symbol of lifeless laughter: a laughter whose power emanates from a source beyond the living. Anarchism will laugh, but its energy will come out of an encounter with death, from the horror of the void. Violence will be the source of true life and the foundation of a new, non-hierarchical authority – a laughing authority that will replace the traditional authority of church and state. 'Ravachol', we are told, 'will perhaps appear one day as a sort of violent Christ, such that the time and milieu he passed through was able to produce'. Both Ravachol and Christ

> wanted to destroy wealth and power, not to seize them. One preached gentleness, spirit of sacrifice and renunciation ... the other preached, through example, revolt against abusive authority, individual initiative against the cowardice of the masses, the claim of the poor to earthly happiness ... They taught the world that neither the ideas of fatherland and society, nor of worship and law, can prevail over the right of man to be happy, whether in this world, as Ravachol said, or in Heaven as Jesus said.[2]

In this chapter I chart some ways in which anarchist urban culture attempted to challenge and reconfigure the life of the city. Anarchism, as has been well documented, proved attractive to many members of the Montmartre artistic and literary avant-garde.[3] Both groups were interested, for example, in exploring shock-effects, hidden symbolisms, and new relationships with nature – as well as the longstanding bohemian practice of attacking established bourgeois moralities and aesthetic codes. George Woodcock suggests that 'It was the anarchist cultivation of independence of mind and of freedom of action and experience

---

2   Ibid. 'Ils voulaient tous les deux, ces démolisseurs du temple, anéantir la richesse et le pouvoir, et non pour s'en emparer. L'un prêcha la douceur, l'esprit de sacrifice et de renoncement, l'affranchissement des entraves politiques par le dédain, en vue de la conquête du royaume céleste; l'autre prêcha par l'exemple la révolte contre l'autorité abusive, l'initiative individuelle contre la lâcheté des masses, la revendication des pauvres au bonheur de la terre. Affranchis de l'égoïsme étroit, ils prirent une plus haute conscience d'eux-mêmes dans l'humanité; partis du principe d'amour, malgré d'apparentes contradictions, ils marchèrent à leur but; avec une volonté héroïque, ils enseignèrent au monde que les idées de patrie et de société, non plus que celles de culte et de Loi ne sauraient prévaloir contre le droit qu'a l'homme d'être heureux, dans ce monde a dit Ravachol, dans le ciel a dit Jésus'.

3   P. Aubery, 'The Anarchism of the Literati of the Symbolist Period', *The French Review* 42(1) (1968), J. Halperin, *Félix Fénéon: Aesthete & Anarchist in Fin-De-Sièecle Paris* (New Haven and London, 1988), J. Hutton, *Neo-Impressionism and the Search for Solid Ground: Art, Science and Anarchism in Fin-De-Siecle France* (1994), R. Roslak, *Neo-Impressionism and Anarchism in Fin-De-Siècle France: Painting, Politics and Landscape* (2007), Anne-Marie Springer, 'Terrorism and Anarchy: Late 19th Century Images of a Political Phenomenon in France', *Art Journal* 38(4) (1979), E. Williams, 'Signs of Anarchy: Aesthetics, Politics, and the Symbolist Critic at the *Mercure de France*, 1890–95', *French Forum* 29(1) (2004).

for its own sake that appealed to the artists and intellectuals'.[4] An important aspect of this relationship, however, has been largely overlooked. This is the link between anarchist *violence* and anarchist *laughter*.[5] Research on fin-de-siècle anarchism has paid curiously little attention to this very striking aspect of anarchism's history. Although we know much about the carnivalesque elements of various anarchist groups since the 1960s, the genealogical roots of this link between anarchism and humour – and its link to debates around revolutionary violence – remain poorly understood.[6] By tracing these links, however, it is possible to gain fresh insight into the spatialities and aesthetics of life, authority and violence in late nineteenth-century Paris. It is such an analysis that I develop in what follows.

## Strange Conjunction: Humour and Violence in Anarchist Urban Culture

As we have seen in earlier chapters, humour was an integral part of the Montmartre avant-garde's attempt to reimagine the life of the city. Central to this experiment was its place-bound nature: Montmartre was imagined as a remote, utopian community, aloof from the sterile financial motives of the city's bourgeois heartlands. Dissatisfaction with this supposed utopia, however, soon started to be voiced. As Montmartre became celebrated as a privileged site of counter-cultural Parisian modernity, and the cabarets artistiques started to achieve considerable financial success, the area began to lose the subcultural capital that it had drawn from claims to authenticity and bohemian, anti-capitalist values. Significant elements of the anarchist community, which had been resurgent since 1880 but by the 1890s had become deeply frustrated by its lack of progress, started to look for ways to communicate their message more effectively in new areas. As a means of doing this, they experimented with the political potential of a distinctively modern technology of communication: dynamite.[7] By the start of the 1890s, the legitimacy

---

4   George Woodcock, *Anarchism. A History of Libertarian Ideas and Movements* (Peterborough, ON, 2004), p. 255.

5   On violence, laughter and affect, see also Maria Hynes, Scott Sharpe and Bob Fagan, 'Laughing with the Yes Men: The Politics of Affirmation', *Continuum* 21(1) (2007).

6   On anarchism, humour and carnival, see, for example, L. M. Bogad, 'Carnivals against Capital: Radical Clowning and the Global Justice Movement', *Social Identities* 16(4) (2010), Paul Routledge, 'Sensuous Solidarities: Emotion, Politics and Performance in the Clandestine Insurgent Rebel Clown Army', *Antipode* 44(2) (2012), Benjamin Shepard, *Play, Creativity and Social Movements: If I Can't Dance, It's Not My Revolution* (London, 2011).

7   As Sarah Cole writes, 'Alfred Nobel's invention of 1866 helped to sweep the world into its modern shape. From the moment of its inception, dynamite violence became an immediate and ever-escalating sensation, with its stunning ability not only to kill and maim, but within seconds to level an entire landscape. The violence of dynamite reverberated in every sensory register as something novel ... from its chemical smell, to its shattering

of the Third Republic, born out of appalling bloodshed, was being challenged by a wave of violence from the radical left. Anarchists, adopting the notorious doctrine of 'propaganda by the deed', unleashed a series of bombings in the bourgeois heartlands of Paris. No longer content with being confined to the marginalized outskirts of Paris, anarchists set out to inscribe the truth within the fabric of the city itself.

In popular stereotypes, the anarchist was a sombre, anti-social, over-sensitive loner. Indeed, the Russian anarchist Sergey Nechayev's notorious pamphlet *Catechism of a Revolutionary* described the anarchist as 'a doomed man' with 'no personal interests, no business affairs, no emotions, no attachments, no property, and no name' and who suppresses 'all the gentle and enervating sentiments of kinship, love, friendship, gratitude, and even honour'.[8] Yet in fact for several years there was a thriving anarchist culture, based largely in Montmartre. Anarchists were not just loners plotting outrages in their bedrooms, but frequented spaces of sociability in which they enjoyed new, modern forms of popular culture. Indeed, they even marketed the thrill and exoticism of anarchist terror to bourgeois visitors.

During the 1880s Montmartre, known for its anti-establishment values and famous as the symbol of the Paris Commune, started to become the home of a number of anarchist cafes and cabarets. One bizarre example of this cross-fertilization of political anarchism and bohemian humour was a cabaret known as the Taverne du Bagne (Jailhouse Tavern). The cabaret, opened by the well-known anarchist Maxime Lisbonne and frequented by anarchists as well as curious bourgeois, was a kind of 'theme bar' that humorously mimicked the conditions of the forced labour camps (the 'bagne') in the South Pacific island of New Caledonia, to where the defeated Communards had been transported in 1871. As Karl Marx's son-in-law Paul Lafargue wrote to Friedrich Engels,

> Lisbonne, professional ham, has had the genial idea of opening a cafe where the doors are barred, where the tables are chained, where all the waiters are dressed as galley slaves, dragging chains … The success has been crazy; one lines up to go drink a bock in the prison of citizen Lisbonne, who makes you pay double on top of it. Members of high society arrive in their carriages, and are happy to hear themselves addressed with 'tu' and to be ill-treated by the prison guards, who use the academic language of the prison to speak to their customers.[9]

---

sound, to its extreme tactile effects … It shattered, exploded, ripped, and tore; it created its own palpable and recognizable form of wreckage; and its employment for radical causes suggested a future with unknowable and potentially frightful contours'. Sarah Cole, 'Dynamite Violence and Literary Culture', *Modernism/Modernity* 16(2) (2009), p. 301.

8    Sergey Nechayev, The Revolutionary Catechism (1869); available at http://www.marxists.org/subject/anarchism/nechayev/catechism.htm, accessed: 6 April 2014.

9    Cited in Richard Sonn, *Anarchism and Cultural Politics in Fin-De-Siècle France* (Lincoln, NE, 1989), p. 130.

As one observer described the scene:

> A shady light fell from the ceiling and a few dirty glass lanterns hung on the pillars. The family son scoffing pancakes at his father's expense, the capitalist who drinks the sweat of the people, found themselves mingling with pimps and dropouts, and as soon as the door opened, each newcomer was received by a volley of insults 'putting them in their place', as we say today. It was a true fountain of manure. ... Waiters looked like crooks with a three day beard. They knew how to walk in their shackles with their large clogs. But sometimes, they studied the shoulders of the lovely ladies, breathing their perfumes to give these doves the shiver of the guillotine.[10]

Like many Montmartre cabarets, a satirical journal was launched to publicize its distinct brand of humour to the general public. 'Between Paris and Montmartre', the journal exclaimed,

> the ex-convict Maxime Lisbonne has just resurrected and resumed the penal colony. This is a daring curiosity that is unique in the history of the fantasies that have made the dear Butte [Montmartre] famous throughout Paris. Staff condemned to serve in the tavern have been picked from former officials, traders, industrial workers, financiers, property owners, priests, brothers, friars, who have suffered their sentences so as to live honestly ... All you who have entered the penal colony – and who, moreover, have got out – thank you for the constant kindness with which you have treated the convicts ... You have helped in the rehabilitation of the fallen, the moralization of rogues.[11]

---

10    Ferdinand Bac, cited in Benoît Noël, *L'Absinthe: Une Fée Franco-Suisse* (Yens sur Morges, 2001), p. 47. 'Un jour louche tombait du plafond et quelques lanternes aux vitres sales étaient accrochées aux piliers. Le fils de famille qui bouffe la galette à papa, le capitaliste qui boit la sueur du people se trouvaient pêle-mêle avec les souteneurs et les raccrocheuse ses et, dès que la porte s'ouvrait, chacun était reçu par des bordés d'injures et "en prenait pour son grade" comme on dit aujourd'hui. C'était une vraie fontaine de purin. ... Les garçons avaient tous des airs de bandits, une barbe de trois jours. Ils savaient marcher dans les entraves avec leurs gros sabots. Mais parfois, ils se penchaient sur les épaules des belles dames en respirant leurs parfums pour donner à ces colombes le frisson de la guillotine'.

11    Maxime Lisbonne, 'La Taverne Du Bagne', *Gazette du Bagne* 1 (1885). 'Entre Paris et Montmartre, ... l'ex-forçat Maxime Lisbonne, vient de ressusciter et de résumer le Bagne. C'est une hardiesse et une curiosité unique dans l'histoire des fantaisies qui ont rendu fameuse la Butte chère aux Parisiens ... Le personnel, attaché au service de la Taverne, a été choisi parmi des anciens Fonctionnaires, Négociants, Industriels, Financiers, Propriétaires, Prêtres, Frères, Ignorantins, qui, ayant subi leur peine qu'à vivre honnêtement ... Vous tous, que êtes entrés au Bagne – et qui, pourtant, en êtes sortis – merci pout la constante bienveillance avec laquelle vous avez traité les forçats ... Vous avez aidé au relèvement des déchus, à la moralisation des dévoyés'.

The anarchist cabaret was a brief sensation, and at one point newspapers could run with the eye-catching headline 'All Paris in the Penal Colony'.[12] Yet, as Michael Wilson notes, the wider social or political purpose of this eccentric provocation to Parisian society was somewhat unclear.

> On the one hand, the Jailhouse Tavern confronted a largely bourgeois public with a particular reality, the traumatic beginnings of the Third Republic. Having a political prisoner serve customers who, while dining, must face depictions of torture and martyrdom does stress the experiential gulf between the conflict's victors and its vanquished. On the other hand, the theatricality of this structured experience must have distanced the customers emotionally from the events on display. Political life becomes a diversion, the meaning of its events inevitably trivialized.[13]

The anarchist cabaret's distinctive brand of humour, that is, seemed to devalue anarchism even as it attacked its bourgeois clients. The cabaret occupied a curious role as *both* a space of political dissent – it was a meeting point for anarchist conspirators (as regular reports from the continual police surveillance carefully noted), and was a venue for rousing talks by the legendary Communard Louise Michel – but also a space for bourgeois diversion and consumption. It betrayed a strange combination of revolutionary zeal with humorous, profit-making diversion. But why did this combination prove so effective in the emergence of a distinctive French anarchist culture? How did such forms of humour help support the practice of anarchist violence?

### Cynicism, Modernity, and the Life of Violence

Recent scholarship on the role of emotions, affect and experience in radical politics has explored the ways in which humour can be used in order to disturb dominant political regimes. In these accounts, the power of humour stems from its excessiveness: its ability to evade, disturb and disrupt established structures of power, organization and representation.[14] Of these accounts, Simon Critchley's

---

12  John Grand-Carteret, *Raphael et Gambrinus, ou L'Art Dans la Brasserie* (Paris, 1886), p. 150.

13  Michael Wilson, 'Portrait of the Artist as a Louis XIII Chair', in Gabriel Weisberg (ed.), *Montmartre and the Making of Mass Culture* (New Brunswick, NJ, 2001), p. 199.

14  See, for example, Kate Epstein and Kurt Iveson, 'Locking Down the City (Well, Not Quite): Apec 2007 and Urban Citizenship in Sydney', *Australian Geographer* 40(3) (2009), David Hammett, 'Resistance, Power and Geopolitics in Zimbabwe' *Area* 43 (2011), Maria Hynes and Scott Sharpe, 'Yea-Saying Laughter', *Parallax* 16(3) (2010), pp. 51–2, Maria Hynes, Scott Sharpe and Bob Fagan, 'Laughing with the Yes Men: The Politics of Affirmation', *Continuum* 21(1) (2007), Anja Kanngieser, *Experimental Politics and the*

account of 'neo-anarchist' ethical subjectivity is arguably the most conceptually powerful. Ethical experience, for Critchley, involves a fidelity to an unfulfillable demand, a demand that the subject acknowledges but can never fully meet. Where the subject might be overcome with melancholy at their inability to meet this demand, humour enables the subject to see itself from the outside and find itself laughably inauthentic. Humour 'recalls us to the modesty and limitedness of the human condition, a limitedness that calls not for tragic-heroic affirmation but comic acknowledgement, not Promethean authenticity but laughable inauthenticity'.[15] This account of humour is an important move in Critchley's argument for a neo-anarchist politics based on infinite responsibility arising in relation to situations of injustice. Critchley draws on the pervasive use of humour in contemporary anarchism's 'new language of civil disobedience', which 'combines street-theatre, festival, performance art and what might be described as forms of non-violent warfare'.[16] These comical tactics 'exemplify the effective forging of horizontal chains of equivalence or collective will formation across diverse and otherwise conflicting protest groups'.

Whilst this theorization of the politics of humour does much to help us understand anarchism's longstanding use of humour as an ethical and political weapon, Critchley's argument that anarchist humour is tied to a logic of pacifism, or what he calls 'non-violent warfare', is problematic. There is no reason, for example, to view the therapeutic effect of humour in managing the burden of excessive ethical demands as being incompatible with the practice of violence. Acknowledging a perceived ethical demand to risk one's life in spreading revolutionary propaganda via acts of violence might require precisely the kind of sublimation through humour that Critchley associates with pacifist politics. Critchley distinguishes the heroic, tragic form of politics from a self-mocking, comedic form of politics. Yet comedy and tragedy are hardly incompatible.[17] Might humour not assist in the creation of the affective resources, the courage and motivation, needed to carry out extremely dangerous acts of revolutionary violence?

---

*Making of Worlds* (Farnham, 2013), Scott Sharpe, Maria Hynes and Robert Fagan, 'Beat Me, Whip Me, Spank Me, Just Make It Right Again: Beyond the Didactic Masochism of Global Resistance', *Fibreculture* 6 (2005), Paul Routledge, 'Sensuous Solidarities: Emotion, Politics and Performance in the Clandestine Insurgent Rebel Clown Army', *Antipode* 44(2) (2012), Marjolein t'Hart and Dennis Bos (eds), *Humour and Social Protest* (Cambridge, 2007), Simon Weaver, 'The "Other" Laughs Back: Humour and Resistance in Anti-Racist Comedy', *Sociology* 44(1) (2010).

15   Simon Critchley, *Infinitely Demanding: Ethics of Commitment, Politics of Resistance* (London, 2007), p. 85.

16   Ibid., 123.

17   On the relationship between tragedy and comedy, see Martin Esslin, *The Theatre of the Absurd* (Garden City, NY: Doubleday, 1961), Louisa Jones, *Sad Clowns and Pale Pierrots: Literature and the Popular Comic Arts in 19th-Century France* (Lexington, KY, 1984), Friedrich Nietzsche, *The Birth of Tragedy*, trans. Shaun Whiteside (London: Penguin, 1993).

This point becomes all the more pressing when we take into account one striking point of intersection between otherwise divergent discourses concerning humour and violence. This intersection comes at the point of an *affirmation of excess* – in particular, an excess of biological life. Theories of revolutionary violence have frequently been attracted by the notion of violence as a form of vital creativity. Hannah Arendt, for example, traces a genealogy of vitalist aestheticization of violence from Bergson and Nietzsche to Sorel, Pareto, Fanon and finally the uprisings of 1968.[18] Common to each was an organicist rendering of politics that framed violent destruction as part of the process of biological growth and creativity. Violence becomes the most authentic source of organic creativity. And in this respect, violence starts bearing a disturbing resemblance to certain organicist theories of the sources and functions of humour. As I explore later in the chapter, nineteenth- and twentieth-century theories of humour saw it as being the most excessive and dynamic expression of human vitality. Both violence and humour were linked by a biopolitical discourse that affirmed their positive role in the proliferation of life, creativity, and excess.

Fully understanding the resonances between violence and humour in nineteenth-century anarchist thought, however, requires unpacking their relationship with truth. Propaganda by the deed was, above all, a practice of militant truth-telling. Indeed, in his lectures on *The Courage of Truth*, Michel Foucault offers some clues as to how such a conjunction of humour, militancy and biological life could productively be theorized within the context of the elaboration of a genealogy of 'arts of living': in particular, to a history of 'parrhesia' or fearless truth-telling. In these lectures, Foucault links the modern revolutionary model of life as a violent, scandalous manifestation of truth to the philosophy of the Cynics in the third and fourth centuries BC.[19] Foucault describes Cynicism as a kind of militancy, 'a militancy in the open ... that is it say, a militancy addressed to absolutely everyone ... which resorts to harsh and drastic means, not so much in order to train people and teach them, as to shake them up and convert them, abruptly'.[20] In this sense, he suggests, the militant movements of the nineteenth century in certain respects echoed the Cynics' techniques for telling scandalous truths, since, like the Cynics, nineteenth-century militants insisted on 'bearing witness by one's life in the form of a style of existence'.[21]

What would it mean to develop a theoretical account of fin-de-siècle cultural politics that is framed in terms of its reactivation of a Cynical ethics of truth-telling? Theorizing anarchist practice in this way, I propose, demands an analysis across three related axes.

---

18   Hannah Arendt, 'On Violence', *Crises of the Republic* (Harmondsworth, 1973).

19   On parrhesia in modern art, see Julian Brigstocke, 'Artistic Parrhesia and the Genealogy of Ethics in Foucault and Benjamin', *Theory, Culture & Society* 30(1) (2013).

20   Michel Foucault, *The Courage of Truth: The Government of Self and Others, Volume Two. Lectures at the Collège de France, 1983–1984* (Basingstoke, 2011), p. 284.

21   Ibid., 184.

First, perhaps the most productive point of contact between the ancient Cynics and modern anarchism comes via the problem of the 'true life'. For the Cynics, speaking truth to power could only be achieved via a highly ascetic art of living. The active transformation of the self was an essential part of the experience of truth. The infamous Cynic Diogenes of Sinope, for example, could speak the truth only via a 'life of scandal' that he insisted was the only *true* life. Diogenes' truths were made possible by a provocatively ascetic life, in which he put his life on display and at risk by means of a range of humorous public interventions, ranging from verbal diatribes to public masturbation. By exposing and endangering his life in this way, Diogenes could describe himself as a true king, an anti-royal king: a person who lives the truly sovereign (that is, anti-sovereign) life. In a Cynical ethics, truth – an embodied, material truth quite unlike the dialogical truth of the Socratics – is interwoven with a style of life, an aesthetics of existence.

Second, and relatedly, viewing anarchist 'propaganda by the deed' as a form of Cynical parrhesia raises the question of *authority*. As Nancy Luxon has argued, an important aspect of ancient parrhesia was the invention of new forms of authority.[22] Parrhesiasts spoke from positions outside those of legitimate authority, and yet had to find ways of making their speech audible, of enabling it to be taken seriously by those whom they advised. A figure such as Diogenes faced the problem of how 'to speak with authority without being authorized to speak'.[23] This issue is particularly pertinent because anarchist theory has been so hostile to the very concept of authority. The paradox of how to speak with authority at the same time as rejecting all forms of authority was, I shall argue, a difficult challenge for anarchists to overcome.

One final aspect of the Cynical truth-telling remains to be drawn out. This concerns the relationship between life and humour in Cynical truth-telling. 'The Cynic', Foucault remarks,

> is a true king; only he is an unrecognized, unknown king who, by the way he lives, by the existence he has chosen, and by the destitution and renunciation to which he exposes himself, deliberately hides himself as king. And in this sense he is the king, but the *king of derision*.[24]

This emphasis on deriding power doubtless partly explains the Cynics' habit of passing down their teachings, not in the form of doctrine, but in 'paigna', jokes and anecdotes which provoked laughter and reported in a few words the gesture,

---

22   Luxon, *Crisis of Authority: Politics, Trust, and Truth-Telling in Freud and Foucault* (Cambridge, 2013).

23   Judith Butler, *Excitable Speech: A Politics of the Performative* (New York; London, 1997), p. 157. See Terry Lovell, 'Resisting with Authority: Historical Specificity, Agency and the Performative Self', *Theory, Culture & Society* 20(1) (2003).

24   Foucault, *The Courage of Truth: The Government of Self and Others, Volume Two. Lectures at the Collège de France, 1983–1984*, p. 278, emphasis added.

retort or attitude of a Cynic in a particular situation.[25] Foucault's remarks on this aspect of Cynicism are very brief, but they are given much more prominence in Peter Sloterdijk's *Critique of Cynical Reason*. Like Foucault, Sloterdijk emphasizes the ways in which the Cynics made the truth dependent on courage, risk and 'cheekiness'.[26] He discerns in Cynicism a 'pantomimic materialism' that is set against the cunning dialectics of the Socratics. Cynicism refutes the language of the philosophers with that of a clown.[27] It 'represents the popular, plebeian rejection of the official culture by means of irony and sarcasm'. Most importantly, it tackles afresh the question of how to *say* the truth, opposing the idealism of the philosophers with a brute materialism, a 'dialogue of flesh and blood'.[28] The Cynic speaks truth with a materialist laughter that explodes with the material, vital energies of urine, faeces, sperm ... and perhaps, we might wonder, blood? It opposes the abstract, dialogical truths of the Socratics with a materialist, embodied, *living* truth. Inserting the story of fin-de-siècle anarchism into a genealogy of ethics, therefore, requires interrogating the new materialities and vitalities of truth that it harnessed.

The remainder of this chapter will diagram the aesthetics of fin-de-siècle anarchist parrhesia by interrogating three related themes: true life, authority and materialist laughter. Theorized as a form of parrhesia, I will argue, 'propaganda by the deed' emerges as an experiment with developing a novel form of courageous truth-telling, one that revives the spirit of the 'king of derision' in order to speak polemical truths about the nature of bourgeois power, authority and truth. In this way, we can begin thinking about anarchist violence and humour as ways of styling an aesthetics of existence, demonstrating anarchists' worthiness of the truth through their willingness to risk their lives in dramatic displays of dissent and derision.

### The City as Medium of Truth

In the fervently positivist intellectual climate of the Third Republic, authority to speak the truth was all but impossible to acquire from outside legitimizing institutions such as scientific laboratories, hospitals, museums, and universities. The medical sciences, in particular, enjoyed huge influence, and parliament was filled with a huge number of doctors.[29] Anarchists had to invent new ways

---

25    Ibid., 208.

26    Peter Sloterdijk, *Critique of Cynical Reason* (Minneapolis, 1987), p. 101. Sloterdijk differentiates ancient 'Kynicism' from modern 'Cynicism', but here I will continue to refer to it as 'Cynicism'.

27    Ibid., 103.

28    Ibid., 104.

29    Jack Ellis, *The Physician-Legislators of France: Medicine and Politics in the Early Third Republic, 1870–1914* (Cambridge, 1990).

of lending authority to their polemical truths. However, this required deploying a new, non-representational *materiality* of truth. Whereas positivist science assumed that truth could be clearly and objectively represented and transmitted to others, anarchists insisted that knowledge could not simply be passed down from 'experts' to the 'ignorant'. Rather, the masses' own immanent capacity to discover the truth of societal exploitation and injustice had to be 'awakened'. The anarchist geographer Peter Kropotkin, for example, made widespread use of this motif of awakening, frequently referring to 'awakening the spirit of revolt' and 'awakening the revolutionary spirit'. For Kropotkin, anarchism's rejection of political representation (that is, parliamentary democracy) necessitated a non-representational practice of truth-telling, one where small acts of revolt would have the effect of *awakening thought* in the masses:

> [N]ot a single revolution has originated in parliaments or in any other representative assembly ... the awakening of the revolutionary spirit always took place in such a manner that, at first, single individuals, deeply moved by the existing state of things, protested against it, one by one. Many perished – 'uselessly', the arm-chair critic would say; but the indifference of society was shaken by these progenitors. The dullest and most narrow-minded people were compelled to reflect, – Why should men, young, sincere, and full of strength, sacrifice their lives in this way? It was impossible to remain indifferent – it was necessary to take a stand, for or against: thought was awakening. Then, little by little, small groups came to be imbued with the same spirit of revolt ... frequently also without any hope of success: simply because the conditions grew unbearable. Not one, or two, or tens, but hundreds of similar revolts have preceded and must precede every revolution.[30]

Propaganda by the deed, then, should be seen as a form of truth-telling that committed itself, not to representing the truth, but to awakening individuals' capacities to *encounter* the truth. This partly explains one thing that has puzzled many critics of anarchism: what message were anarchist acts of violence intended to convey? Anarchist violence resembled, as Howard Lay suggests, 'a primal utterance, a solitary "!" reduced to its bare bones without a word, a phrase, or a sentence to precede it'.[31] It appeared to speak a truth with no content, leading to the common presumption that anarchists were simply nihilists, hellbent on destruction for its own sake. In fact, however, propaganda by the deed was intended to act as a shock that would awaken the truth. Through violence, that is, anarchists sought a distinctive form of truth-telling wherein truth, rather than being represented, would spontaneously grow out of the shock of sheer affect. Anarchists, rejecting

---

30   Peter Kropotkin, 'Modern Science and Anarchism', in Roger Baldwin (ed.), *Kropotkin's Revolutionary Pamphlets* (New York, 1970).

31   Howard Lay, '"Beau Geste!" (on the Readability of Terrorism)', *Yale French Studies* 101 (2001), p. 80.

the scientific institutionalization and bureaucratization of truth, sought to invent new languages of non-representational, affective truth-telling.

In order to find such a language, anarchists looked back with nostalgia to 'primitive' societies.[32] One humorous anarchist pamphlet collected in the police archives, for example, imagined a 'primitive' man, magically transported to the present day, in conversation with four men of modernity: a mine worker, a factory labourer, an agricultural worker and an office clerk.[33] As the group stand in front of symbols of French modernity such as the Eiffel Tower, the primitive man, stupefied at seeing their unhealthy physique, asks them a series of questions: 'Why these black marks all over your white face? Why are you so thin and enfeebled? Why do you have no hair or teeth? Why are you so broken and tired?' The pamphlet goes on to lyrically describe the happy life of primitive man, with his easy life of hunting, dancing and abundant sex, not to mention his average life expectancy of 120 years. 'And now, Sociologists', the article concludes, 'compare this with the life of the Proletarian in civilization!'[34] Pre-modern man, the article emphasizes with gentle humour, enjoyed a vigorous vitality that has withered in the modern age.

One feature of primitive societies that anarchists found compelling was the non-representational form of language that they imagined to have been practiced. This aspect of anarchist thought can be traced back to Jean-Jacques Rousseau's 1755 *Essay on the Origin of Languages*, where Rousseau had suggested that language originated in an attempt to strengthen the emotional power of primitive people's grunts and cries to each other.[35] Language originated as an exchange of passions.[36] For this reason, 'at first only poetry was spoken'.[37] The first words weren't abstractions or signs, but direct affects. However, as language became more oriented towards practical needs with the growth of civilization, language became more abstract and prosaic. It started to defer experience, rather than fulfilling it. One effect of modernity, in this Romanticist discourse, is that language lost its unmediated relationship with the world. Language as a circulation of affects gave way to an artificial system of representation. Language lost its bond with embodied experience.

---

32    On the role of nostalgia in radical politics, see Alastair Bonnett, 'The Nostalgias of Situationist Subversion', *Theory, Culture & Society* 23(5) (2006), Alastair Bonnett, 'The Dilemmas of Radical Nostalgia in British Pyschogeography', *Theory, Culture & Society* 26(1) (2009).

33    Anon., 'L'Etat Naturel et la Part du Proletaire dans la Civilisation', (Archives de la Prefecture de la Police, n.d.).

34    Ibid. 'Et maintenant, Sociologues, comparez à cela, l'existence du Prolétaire dans la Civilisation!'

35    Jean-Jacques Rousseau, *Essay on the Origin of Languages and Writings Related to Music* (Hanover, NH, 1998).

36    On Rousseau and the myth of the noble savage, see Terry Ellingson, *The Myth of the Noble Savage* (Berkeley, CA, 2001).

37    Rousseau, *Essay on the Origin of Languages and Writings Related to Music*.

Significant elements of this dream of speaking the truth by reuniting language with affective experience are clearly discernible in the practice of propaganda by the deed. Indeed, anarchists frequently emphasized the *symbolic* dimensions of their acts of revolt. Many saw their distinctive form of truth-telling to be continuous with the non-representational modes of communication that were simultaneously being developed by the Symbolist literary avant-garde.[38] Symbolism was an artistic movement based on a fascination with exploring the hidden, mysterious, primeval energies lurking beneath the everyday order of things. The poetry of Mallarmé, for example, unravelled the structures of linguistic representation, abandoning conventional narrative and description whilst seeking new forms of meaning in the flights of association and imagination beyond the word's 'practical' use of signification.[39] Mallarmé paid close attention to the embodied aspects of language, evoking new meanings through novel use of textures of sound, rhythm and symbol. His work was taken up by a school of followers who attempted to create forms of symbolic poetry which, as Jean Moréas put it, 'clothe the Idea in a sensible form which, nonetheless, will not be an end in itself, but which, while serving to express the Idea, remains the subject'.[40] Symbolists strove above all else to create a newly *embodied*, non-representational form of language.

This impulse to 'dress the Idea in sensible form' is also clearly discernible in anarchist political culture. Anticipating the vitalist theory of violence expounded a decade or so later by the anarcho-syndicalist Georges Sorel, fin-de-siècle anarchists emphasized the ways in which violence made ideas more *vital*, more living.[41] 'The idea', wrote Georges Brousse, 'will not appear on paper, nor in a newspaper, nor in a painting; it will not be sculpted in marble, nor carved in stone, nor cast in bronze: it will walk, alive, in flesh and bone, before the people. The people will hail it as it passes'.[42] Through violence, that is, ideas would *come alive*. Rather than representing pre-formed ideas, the aim of anarchist violence was to create the shock to thought that would awaken the truth and make it multiply and differentiate itself, resonating throughout society, evoking new passions in

---

38    On anarchism and Symbolism, see Aubery, 'The Anarchism of the Literati of the Symbolist Period', Erin Williams Hyman, 'Theatrical Terror: Attentats and Symbolist Spectacle', *The Comparatist* 29 (2005), Richard Shryock, 'Becoming Political: Symbolist Literature and the Third Republic', *Nineteenth Century French Studies* 33(3 and 4) (1958), Williams, 'Signs of Anarchy: Aesthetics, Politics, and the Symbolist Critic at the *Mercure de France*, 1890–95'.

39    On the echo of Rousseau in Symbolists' approach to language, see Chapter 1 of Richard Candida Smith, *Mallarmé's Children: Symbolism and the Renewal of Experience* (Berkeley, CA, 1999).

40    Jean Moréas, 'Le Symbolisme', *Le Figaro, Supplément Littéraire*, 18 September 1886. '[L]a poésie symbolique cherche à vêtir l'Idée d'une forme sensible qui, néanmoins, ne serait pas son but à elle-même, mais qui, tout en servant à exprimer l'Idée, demeurerait sujette'.

41    Georges Sorel, *Reflections on Violence*, trans. Jeremy Jennings (Cambridge, 1999).

42    Brousse, cited in Hyman, 'Theatrical Terror: Attentats and Symbolist Spectacle'.

those whom it touched. In this way, violence would also create a dramatically new *spatiality* of truth. Truth would no longer circulate around a few cafes in the working class peripheries and amongst the journals of the already initiated. Rather, bombs would sow seeds of truth in the very heartlands of bourgeois Paris. Through violence, anarchist truth-telling could assert a new spatiality of life itself, via a living truth that would spread itself around the areas of the city where it would be most disruptive. The urban fabric itself – not mere pamphlets, posters and newspapers – would become the new medium of revolutionary truth.

So propaganda by the deed can be viewed as a spatial practice that established a new, affective relationship between truth and life. Through violence, truth itself could come alive. In this way, anarchism established a newly materialized practice of truth, one where truth acquired a new spatiality, becoming present within the city's cafes, hotels, boulevards and apartment blocks as a growing sense of disturbance and unease, a premonition of revolutionary change. The non-representational nature of violence, the semiotic emptiness that gave it such primeval affective power, aimed to provide a shock to thought that would enliven the truth, giving it the power to spread and multiply – initiating, in the words of Barrucand cited earlier, a 'healthy contagion'.

Yet other possibilities than violence were open to anarchists keen to invent a non-representational style of political truth-telling. Indeed, many turned instead to more literary and artistic forms of anarchism.[43] Why, then, did some anarchists see propaganda by the deed as the only effective solution to their problem? The answer to this question, perhaps, has something to do with the fact that the anarchist reactivation of the spirit of the Cynic 'king of derision' needed not only to speak the truth, but to speak the truth *with authority*.

## Immanent Authority

For the Cynics, the ability to make speech acts authoritative was linked to the parrhesiast's courage: his willingness to risk his life in order to demonstrate the power and authenticity of his speech. It is this courage that enabled truths to be spoken with authority despite coming from outside the dominant institutional or transcendent sources of authority. A similar theme is clearly discernible in anarchist theories of propaganda by the deed. In this section I wish to argue that anarchists' acts of individual revolt, through their courageous confrontation with death via the guillotine, were intended to create a form of *immanent authority* for their political speech acts.[44]

---

43   See David Weir, *Anarchy & Culture: The Aesthetic Politics of Modernism* (Amherst, 1997).

44   On 'immanent authority', see also Julian Brigstocke, 'Immanent Authority and the Performance of Community in Late Nineteenth Century Montmartre', *Journal of Political Power* 6(1) (2013).

In Barrucand's celebration of anarchist violence (cited in the introduction to this chapter), he alludes to the 'legendary authority' of Ravachol's severed head. The phrase seems jarring; anarchism after all, is defined by its unremitting hostility to authority. Yet fin-de-siècle anarchism actually did have a positive theory of authority embedded within in it – and this theory of authority helps explain the role of violence in anarchist parrhesia.

'Authority', Mikhail Bakunin wrote in *God and State*, is 'a word and a thing which we detest with all our heart'.[45] Yet, he emphasized, anarchism does not reject all authority, only *external* authority.[46] It is happy, for example, to obey 'the inevitable power of the natural laws which manifest themselves in the necessary concatenation and succession of phenomena in the physical and social worlds'.[47] Slavery to natural laws is no slavery at all, because such laws 'constitute the basis and fundamental conditions of our existence; they envelop us, penetrate us, regulate all our movements, thoughts, and acts'. Indeed, Bakunin went on, anarchism is really about rejecting *singular*, absolute authority: 'I do not content myself with consulting a single authority in any special branch; I consult several; I compare their opinions, and choose that which seems to me the soundest. But I recognise no infallible authority … [and] I have no absolute faith in any person'.[48] This means that 'there is no fixed and constant authority, but a continual exchange of mutual, temporary, and above all, voluntary authority and subordination'. This means that anarchism, whilst being wholly opposed to the singular authorities of God and Law, is compatible with voluntary submission to the authority of nature – and hence, the authority of the spokesman of nature, science. 'We recognize, then, the absolute authority of science … [which is] legitimate because rational and in harmony with human liberty'.[49] Bakunin's vision for anarchism, far from rejecting authority wholesale, attempted to mobilize those forms of authority that were immanent to natural life. Life – in all its plurality, creativity, and heterogeneity – was to be the ultimate source of authority in an anarchist society.

For anarchist truth to become authoritative, therefore, it had to harness the authority of natural life. In this respect, anarchist theory evidently participated in wider biopolitical discourses that established life, health and natural vitality

---

45 Mikhail Bakunin, *God and the State* (New York, 1970).

46 Despite the centrality of the concept of authority to anarchist theory, it can sometimes be frustratingly imprecise concerning its use of the term. Often it conflates the rejection of *all* authority with the rejection of *artificial* authority. Thus Saul Newman, for example, can argue in the same article both that what unites anarchism is 'fundamental rejection and critique of political authority in all its forms'; and also that 'Anarchism may be understood as a struggle between natural authority and artificial authority'. These two positions are entirely distinct, however. The latter position allows for forms of natural political authority – precisely the kind of authority I explore in this chapter. See Saul Newman, 'Anarchism and the Politics of Ressentiment', *Theory & Event* 4(3) (2000).

47 Bakunin, *God and the State*, p. 29.

48 Ibid.

49 Ibid., 34.

as cardinal reference points for evaluating social and political practices and structures, viewing life as 'the supreme standard and the highest good to which everything else is referred'.[50] Whilst it differed strongly in its interpretation of the laws governing natural life (most importantly, stressing the role of co-operation over competition in evolutionary processes), anarchism did little to challenge the escalating authority of biological discourses themselves.[51] This 'biopolitical authority' operated at two poles.[52] On the one hand, it operated at the level of objective knowledge, with anarchist intellectuals such as Kropotkin and Reclus engaging in hugely ambitious projects to synthesize the human, natural and mineral worlds through their vast universal geographies. On the other hand, there was a growing movement towards locating the authority of life, not just in objective knowledge, but in embodied experience. Experience, as the movement of life, was to become its own authority.

Evidence of this can be seen in Barrucand's allusion to a 'new morality without obligation or sanction' in his essay on Ravachol. This is a clear reference to a book titled *Sketch of Morality Independent of Obligation or Sanction* by the sociologist-philosopher Jean-Marie Guyau. Guyau, a now largely overlooked figure in the history of the social sciences, was widely celebrated in the late 1880s. He had an important influence on thinkers such as Emile Durkheim (who took his theory of anomie from Guyau's *The Non-Religion of the Future*, inverting it from a positive to a negative phenomenon); Friedrich Nietzsche (who made copious notes on Guyau's *Sketch of Morality*); and Peter Kropotkin (who celebrated Guyau's work in his uncompleted *Ethics*, and described Guyau as being 'unconsciously anarchist' in his thinking).[53]

Guyau argued that the only legitimate form of obligation or authority comes, not from any external source, but from a force that is immanent to life and lived experience. Guyau's work explored the potential for developing a purely biological theory of obligation.[54] Drawing on evolutionary theory, as well as on

50    Johanna Oksala, 'Violence and the Biopolitics of Modernity', *Foucault Studies* 10 (2010), p. 25.

51    On anarchism and evolutionary theory, see Peter Kropotkin, *Mutual Aid: A Factor of Evolution* (1987), Peter Singer, *A Darwinian Left: Politics, Evolution and Cooperation* (London, 1999).

52    On biopolitical authority, see also Claire Blencowe, 'Biopolitical Authority, Objectivity and the Groundwork of Modern Citizenship', *Journal of Political Power* 6(1) (2013).

53    Keith Ansell-Pearson, 'Free Spirits and Free Thinkers: Nietzsche and Guyau on the Future of Morality', in Jeffrey Metzger (ed.), *Nietzsche, Nihilism and the Philosophy of the Future* (London, 2009), Peter Kropotkin, *Ethics: Origin and Development*, trans. Louis S. Friedland and Joseph R. Piroshnikoff (New York, 1924), Kropotkin, 'Anarchist Morality', p. 108, Marcho Orru, 'The Ethics of Anomie: Jean Marie Guyau and Émile Durkheim', *The British Journal of Sociology* 34(4) (1983).

54    Jean-Marie Guyau, *Education and Heredity: A Study in Sociology*, trans. W. J. Greenstreet (1891), Jean-Marie Guyau, *A Sketch of Morality Independent of Obligation*

Alfred Fouillée's call for a theory of moral obligation that makes the moral ideal 'immanent' to experience, Guyau dismissed any attempt to justify moral obligation from a metaphysical or religious point of view.[55] The feeling of obligation or duty, he insisted, is precisely a *feeling* – and a feeling, not of necessity or compulsion, but of *power*, where power is simply an abstraction of organic life. In other words, the feeling of obligation is a feeling of inner life that is forceful enough to compel action. 'To act is to live; to increase action is to increase the fire of inward life ... life is thus, in fact, the primitive and universal form of every good thing which is desired'.[56] All action is a form of growth, an intensification of life, and for this reason, 'it is from life that we will demand the principle of morality'.[57] We act, not because we choose to, but because life obliges us to.

Guyau's anarchic (or what he called 'anomic') conception of morality made the experience of life the key source of authority. The only authority that should be obeyed, in this ethical framework, is the authority of experience – for it is here that true life asserts itself. Strikingly, Guyau linked this new authority to the experience of risk and peril. Just as Ferdinand Tönnies, in his seminal 1887 work *Community and Society*, highlighted the importance of strength and courage in the production of authority, Guyau suggested that the highest intensification of life – the *truest* life – is found in the experience of peril.[58] Indeed, the experience of risk is 'life itself raised into sublimity'.[59] To confront risk is to place one's body in a new, hostile milieu; by confronting risk, individuals force themselves to adapt to this environment and thereby grow stronger and more adaptable. In turn, the environment itself adapts to the new, stronger forms of life within it. It is by stepping outside the usual environments of life and confronting peril that life becomes most creative, dynamic and intense. True life, that is to say, comes from the experience of peril. In courageously risking one's life, one's actions can harness the authority *of* life.

It is easy to see why Guyau's account of the authority of embodied experience should have proved so attractive to fin-de-siècle anarchists. It created a theoretical underpinning for the individual acts of revolt that anarchists were convinced were necessary in order to revitalize and rematerialize truth. By risking (or, indeed, sacrificing) their lives in speaking truth to power, anarchists could claim to be living a true, authentically *vital* life. Through a certain stylization of their life they could lend added authority to their speech acts, creating an epistemic ecology in which

---

or *Sanction*, trans. G. Kapteyn (London, 1898), Jean-Marie Guyau, *The Non-Religion of the Future: A Sociological Study* (New York, 1962).

55   On Fouillée, see Robert Good, 'The Philosophy and Social Thought of Alfred Fouillée' (PhD thesis, McGill University, 1993).

56   Guyau, *A Sketch of Morality Independent of Obligation or Sanction*, pp. 80–81.

57   Ibid., 70.

58   Ferdinand Tönnies, *Community and Civil Society*, trans. Jose Harris and Margaret Hollis (Cambridge, 2001), p. 30.

59   Guyau, *A Sketch of Morality Independent of Obligation or Sanction*, p. 125.

their truths would find the conditions for vigorous organic growth. This vitality, born of the experience of risk and the courage of truth, was to create the authority for forms of non-representational truth-telling that would create new forms of autonomous, organically self-forming and rapidly fructifying collective revolt. Revolt was the paradigmatic way by which the life of the city could overreach itself and creatively transform itself. Through courageous acts of violence, anarchist truth could mobilize the highest authority of all: the authority of life itself.

Yet the contradictions inherent in critiquing biopolitical rationalities by pointing to a *truer* life, rather than constructing a political rationality that refused the modern imperative to take life as the highest value, meant that in practice, anarchist violence struggled to gain the authority it sought for its polemical truth claims. The authority of anarchism's violent speech acts was easily undermined by discourses that portrayed anarchists as abnormal, pathologically ill criminals. For example, the celebrated anthropological criminologist Cesare Lombroso compared the outbreak of anarchist violence with an outbreak of cholera, and asserted, with the full authority of medical science behind him, that 'it is easy to see that the anarchist movement is made up for the most part ... of criminals and madmen, and sometimes both together'.[60] One exception to this, he wrote, was the bomber Auguste Vaillant, whom he instead diagnosed as a hysteric, a product of degenerate parents and an unhappy upbringing. Interestingly, Lombroso used this diagnosis to attack directly any claim of *courage* on Vaillant's part. He regarded him as a particular kind of hysteric, a 'charitable' hysteric. 'The charitable hysteric', he writes,

> is likely to accomplish feats of courage that are cited and repeated or which even become legendary. If a fire broke out, she could be a quite superior presence of mind, give excellent advice, shelter personal effects and livestock, or plunge into the middle of the flames to save a cripple, an old man or a child.[61]

Crucially, however, this courage is a *pathological* form of courage. Charitable hysterics 'play the role of virtue pathologically, and everyone falls for the trick'.[62] Even the anarchist's courage is not evidence of vitality and vigour, but of degeneracy and disease.

---

60   César Lombroso, 'L'Anarchie et Ses Héros', *La Revue des Revues*, 15 February 1894, p. 270. 'Il est facile dès lors de comprendre que les fauteurs de l'anarchie soient composés ... en majeure partie de criminels et de fous et quelquefois des uns et des autres ensembles'.

61   Ibid., 273. 'L'hystérique charitable est susceptible d'accomplir des trait de courage qui sont cites et répétés ou qui deviennent même légendaires. Qu'un incendie éclate et elle pourra faire preuve d'une présence d'esprit tout à fait supérieure, donnera des conseils excellents, fera mettre à l'abri les objets mobiliers et les bestiaux ou se précipitera au milieu des flammes pour sauver un infirme, un vieillard ou un enfant'.

62   Ibid., 274. 'Elles jouent pathologiquement le rôle de la vertu, et tout le monde s'y laisse prendre'.

This discursive reimagining of anarchist propaganda by the deed in newspapers and scientific publications indicates some of the limitations of the anarchist theory of communication. Whilst the act of violence itself was indeed an excessive, non-representational form of communication, a 'solitary "!"', this excessive event could quickly be re-appropriated within hegemonic scientific discourses.[63] The meaninglessness of the events only made it easier for new meanings to be imposed on them by dominant epistemological authorities.

By itself, then, violence could not do the work needed of it: to create a non-representational medium for voicing militant truths that would gain authority from the courage of the truth tellers. Anarchists could too easily be portrayed as weak, sickly, cowardly individuals – 'monstrosities' who threatened a healthy social body. It is in this context, I will argue, that we should view the use of humour in anarchist culture. In humour, anarchists could lay claim to a form of experience that was much less problematically healthy and vigorous.

**A True Fountain of Manure**

Violence was one technique for making anarchist truth-telling authoritative by harnessing the non-representational, excessive, affective qualities of violence. The authority of these speech acts, however, could easily be undermined by pointing to the madness, illness or congenital criminality of its perpetrators. In humour, by contrast, I wish to argue, anarchism found an alternative way of deploying affect that could exert a different claim to vitality and creativity.

In Hannah Arendt's landmark study of political authority, she argues that '[t]he greatest enemy of authority … is contempt, and the surest way to undermine it is laughter'.[64] Laughter is a powerful way to attack authority, since it undermines respect for those who hold positions of authority. This link between authority and humour offers one explanation for fin-de-siècle anarchism's fondness for humorous urban culture. Contemptuous of all external authority, anarchists recognized that humour was one powerful means of undermining authority.

Certain wings of anarchism, as we have seen, were fascinated with the possibilities of replacing external, transcendent authority with the immanent forms of authority that were internal to life itself, and the embodied experiences through which life expresses itself. If violence offered one aesthetic structure for achieving this, humour offered a parallel one, since, as explored in Chapter Five, it was humour that had recently emerged in many influential evolutionary biological discourses as being the most excessive and dynamic expression of human vitality. The work of Alexander Bain and Herbert Spencer, for example – as translated and popularized by the founder of French experimental psychology,

---

63   See Lay, '"Beau Geste!" (on the Readability of Terrorism)'.
64   Hannah Arendt, 'What Is Authority?', *Between Past and Future: Six Exercises in Political Thought* (New York, 1961).

Théodule Ribot – emphasized the role of laughter in dispersing excessive energy.[65] Bain had argued that laughter represented an increase of vitality and a 'heightening of the powers of life'.[66] Laughter, he suggested, accompanies a release from constraint: in other words, it is an experience of *freedom*.[67] Spencer, similarly, argued that laughter is the 'result of an uncontrolled discharge of energy': it is the physiological response to an excess of subjective feeling.[68] Charles Darwin, meanwhile, had used laughter to demonstrate the commonalities between humans and monkeys.[69] Laughter, then, was acquiring a prominent place in emerging evolutionary discourses as an expression of biological excess and affective energy.

These emerging knowledges of laughter as an authentic expression of biological vitality, I propose, may help to explain the affinity that anarchists felt to the humorous culture of Montmartre's cafes and cabarets. Certainly there was an unmistakeably carnivalesque element to much popular anarchist culture – a riotous fusion of revolutionary terror with explosive laughter and furious dancing. This sense is captured well by the popular song written in the name of 'Lady Dynamite':

> Our fathers once danced
> To the sound of the cannons of the past!
> Now this tragic dance
> Requires stronger music.
> Let's dynamite, let's dynamite!
> Lady Dynamite, let's dance fast!
> Let's dance and sing!
> Lady Dynamite, let's dance fast!
> Let's dance and sing, and dynamite![70]

A similar coming together of violence, dance and laughter was expressed in the infamous and widely sung ditty 'La Ravachole':

> In the great city of Paris,
> There are well fed bourgeois
> And also the destitute

---

65    Théodule Ribot, *La Psychologie Anglaise Contemporaine*. (1870). See Michael Billig, *Laughter and Ridicule: Towards a Social Critique of Humour* (London, 2005), p. 104.

66    Alexander Bain, *The Senses and the Intellect* (London, 1855), p. 251.

67    Alexander Bain, *The Emotions and the Will* (London, 1859).

68    Herbert Spencer, 'The Physiology of Laughter', *Essays: Scientific, Political and Speculative* (New York, 1864).

69    Charles Darwin, *The Expression of the Emotions in Man and Animals* (London, 1872).

70    Cited in John Merriman, *The Dynamite Club: How a Bombing in Fin-De-Siècle Paris Ignited the Age of Modern Terror* (London, 2009), p. 85.

Who have empty stomachs.
But they have long teeth.
Long live the sound,
They have long teeth,
Long live the sound,
Of explosion!
Let's dance La Ravachole,
Long live the sound
Let's dance La Ravachole,
Long live the sound,
Of explosion!
Ah, it's coming, it's coming,
All bourgeois will have a taste of the bomb!
It's coming, it's coming,
All bourgeois will be made to jump,
We'll make them jump![71]

Songs such as these testify to the centrality of a cultural politics based on a celebration of *energy* in all its forms to the anarchist movement. Laughter and dancing were widely regarded as expressions of animalistic vital urges (and associated with creative maladies such as hysteria), and in this context became associated with an anarchic liberation of natural energy and vitality.[72] Anarchist culture now added dynamite to laughter and dance as means of liberating

---

71   'Dans la grand'ville de Paris,
      Il y a des bourgeois bien nourris;
      Il y a aussi des miséreux
      Qui ont le ventre bien creux.
      Ceux-là ont les dents longues,
      Vive le son, vive le son,
      Ceux-là ont les dents longues,
      Vive le son
      D'l'explosion!
      Dansons la Ravachole,
      Vive le son, vive le son,
      Dansons la Ravachole
      Vive le son de l'explosion!
      Ah, ça ira, ça ira, ça ira,
      Tous les bourgeois gout'ront d'la bombe!
      Ah! ça ira, ça ira, ça ira,
      Tous les bourgeois on les saut'ra …
      On les saut'ra!'
      For a discussion of the song and its variations, see Chapter 5 of Alexander McKinley, *Illegitimate Children of the Enlightenment: Anarchists and the French Revolution, 1880–1914* (New York, 2008).

72   See Chapter 5.

or reconstructing the life of the city. All became expressions of what Baudelaire had once referred to as life's 'luminous explosion in space'.[73]

Montmartre, however, as the spiritual home of anarchism, was humorously framed as being the only area of the city in which Parisians could count themselves safe. After Ravachol's bombings in 1892, the anarchist Maxime Lisbonne spread placards across Montmartre, offering insurance policies against dynamite explosions, contracted by a new Committee of Public Safety.[74] Female property owners could avoid the need for such policies simply by marrying an anarchist. The proclamation was accompanied by the comment, 'One laughs at everything on the Butte'. In 1894 he advertised another cabaret, the Concert Lisbonne, as 'the sole Concert sheltered from the bombs'.

Humour offered anarchists a novel kind of aesthetic intervention that might reverse hierarchies, challenge norms, and disturb power structures. It harnessed a powerful form of non-representational affective energy – a surplus that could not be captured. Using humour was a way of demonstrating the health and vitality of anarchist ideas. In doing so, they could lend greater legitimacy to their claim that violence, far from being a regressive practice that would contribute to the degeneration of the population, was a form of creative destruction, a temporary reversal that would make possible a purer form of human evolution. This more advanced form of human sociality would be based on co-operation and peace rather than competition and violence.

It is here that we may return to the image of Ravachol's laughing corpse. The laughter on Ravachol's lips that was an important motif in the anarchist hagiography of Ravachol, we might surmise, comes from the *vitality* within death that it signifies. Anarchism may deal in death, but it is a death that is invested with a laughing vitality that will help restore human society to its natural, uncorrupted state of being. By laughing, death can become life, violence can become creative, and the marginalized can assert their superiority over their rulers.

In a curious way, the tactic of anarchist derision was unwittingly valorized in the criminologist Lombroso's own text on anarchism. The bomber Vaillant, he argued, must not be executed. Harsh penalties against anarchists merely create more martyrs and reinforce the cycle of violence. 'A more radical measure, above all in France', he argued, 'would be to cover them in *ridicule*. Martyrs are venerated; but never the mad'.[75] Derisory laughter, too, could be used as a weapon against anarchists. It is easy to laugh at the mad. Laughter, the pre-eminent

---

73    Charles Baudelaire, 'The Painter of Modern Life', in Jonathan Mayne (ed.), *The Painter of Modern Life, and Other Essays* (London, 1964).

74    Richard Sonn, 'Marginality and Transgression: Anarchy's Subversive Allure', in Gabriel Weisberg (ed.), *Montmartre and the Making of Mass Culture* (New Brunswick, NJ, 2001), p. 130.

75    'Une mesure plus radicale, surtout en France, serait de les couvrir de ridicule. Les martyrs sont vénérés, jamais les fous'. Lombroso, 'L'anarchie et ses Héros'.

anti-authoritarian weapon, could be used to undermine the authority of anarchism, that most anti-authoritarian of political movements.

Whilst anarchist humour did little to convince the bourgeois audience whom it attacked, it did set an agenda for a continuing affinity between humour and radical political action in coming decades. André Breton's *Anthologie de Humour Noir*, for example, is filled with writers who described themselves as anarchists.[76] Alain Faudemay lists several reasons for humourists' affinity with anarchism: their cynical outlook; their fondness for the macabre; their scepticism towards social values; their weakness for mystification; their interest in fragmentary, anti-authoritarian structures.[77] In this chapter, I have indicated one important further reason: their interest in discovering new forms of disruptive, creative social vitality in the affective and discursive excesses of embodied humour.

## Conclusion

In 1893 and 1896, a period in which anarchist bombs were regularly terrorizing the city of Paris, two theatrical productions that would have a formative influence on the development of the twentieth-century avant-garde were premiered at the Théâtre de l'Œuvre, at the foot of Montmartre. The productions were of Henrik Ibsen's *An Enemy of the People* and Alfred Jarry's *King Ubu*. Both were described as 'attentats' (terrorist attacks), and their authors labelled 'anarchists of art' who were 'exercising a veritable terror over the public'.[78] Both were comedies, or at least contained comic elements.[79] Both also deployed distinctively biopolitical themes. *An Enemy of the People*, for example, dramatizes the events in a town that is growing wealthy from a new health spa, which is in fact poisoning the tourists who visit.[80] When the hero, Dr Stockman, discovers that the water is polluted and alerts the community, he finds the entire town determined to cover up this threat to its livelihood. The community declares the water safe and Stockman is eventually ostracized and denounced as a lunatic. 'Health', here, is poisonous. Moreover, the only person to alert the community to this health threat is again denounced in the name of health. The 'life of the city' is revealed as being deeply corrupt and unhealthy. The oppressiveness of community is opposed to heroism of individual life: 'the strongest man in the world is he who is most alone', as the final line of play has it.

---

76   Andre Breton, *Anthology of Black Humor* (San Francisco, 1997).
77   Alain Faudemay, 'L'humour et L'Anarchisme: Quelques Indices d'une Convergence Possible', *Revue d'histoire littéraire de la France* 99(3) (1999).
78   Henri Fouquier, 'Review of Ubu Roi', *Le Figaro*, December 11 1896. Cited in Erin Williams Hyman, 'Theatrical Terror: Attentats and Symbolist Spectacle', *The Comparatist* 29 (2005).
79   For a reading of *An Enemy of the People* as a comedy, see Harold Knutson, '*An Enemy of the People*: Ibsen's Reluctant Comedy', *Comparative Drama* 27(2) (1993).
80   Henrik Ibsen, *An Enemy of the People*.

Jarry's iconoclastic *Ubu Roi*, whose assault on all theatrical conventions is often considered to mark the birth of avant-garde theatre, also explores biopolitics themes.[81] The anti-hero of Jarry's iconoclastic play is a grotesque caricature of the upper-middle class. Lazy, violent, cowardly, avaricious, stupid and greedy, Père Ubu is obsessed with the ultimate symbol of filth: shit. The play's scandalous first word, 'merdre!' (an invented word often translated as 'pschitt'), metaphorically threw shit into the audience's faces. Jarry uses shit as a symbolic device through which to announce the utter corruption of bourgeois modernity. Modernity, obsessed with health and cleanliness, has created only filth and disease.

These performances became new symbols of a turning point in Montmartre urban culture. Towards the end of the nineteenth century, the strong links between popular culture, avant-garde culture and anarchist politics in Montmartre began to loosen. Much of the force of anarchist politics, argues David Weir, was incorporated into the cultural freedoms of modernist art.[82] Anarchist politics moved towards less violent, less humorous and more constructive anarcho-syndicalist politics. Popular culture began to lose its ties to radical urban culture as it became ever more commodified. Montmartre, whilst remaining a focus of artistic experimentation for the next 20 years, lost the aura of authenticity that it had possessed during the 1880s.

In this chapter I have constructed a theoretical diagram that captures some of the intersections of violence and laughter in fin-de-siècle anarchism. Anarchism, we have seen, had to tackle an insurmountable series of contradictions as it was feeling its way towards a viable political alternative to the Third Republic. Its curious embrace of humour and violence, I have argued, can be explained by the affective structure that was common to each. Violence offered a way of disrupting the established order of visibilities and invisibilities, planting a newly materialized truth within the fabric of the city itself. Through dynamite, a new kind of atmosphere could be secreted throughout the city's bourgeois spaces, an atmosphere filled with the potentiality of imminent violence. Moreover, the courage in the face of death that such violence demanded could serve as a demonstration of anarchists' vitality and commitment. Yet dominant discursive regimes could easily undermine anarchism's claim to a more authentic vitality, portraying anarchism instead as a symptom of degeneration, irrationality and feminine over-sensitivity. Anarchism's use of laughter, by contrast, offered a more unproblematically vital form of embodied experience. Through laughter, it could demonstrate its proximity to true nature.

Yet the anarchist movement was inhibited by its inability to escape the biopolitical discourses it criticized so ferociously. Rather than performing a militant refusal to take life as the key source of value in modern society, anarchism instead attempted to discover a *truer* life, exposing bourgeois life as a lie and pointing the way to a more authentically natural way of living. Yet anarchism's complicity with

---

81    Alfred Jarry, *Ubu Roi*.
82    Weir, *Anarchy & Culture: The Aesthetic Politics of Modernism*.

the liberal quest to secure and maximize the life of the population meant its claim to a truer life could always be undermined by authoritative scientific discourses. In laying claim to knowledge of true life, it could not compete with institutionally legitimized knowledge claims concerning the nature of healthy life. Moreover, this complicity with biopolitical discourses carried the danger of lapsing into the kind of deeply disturbing proto-fascistic anarchism associated with later Bergsonist avant-garde anarchist groups such as *Action d'Art*.[83] In adopting the discourse of life, anarchism tied itself so closely to dominant liberal regimes of government that it failed to offer an alternative that would be powerful enough to attract widespread support or initiate the 'healthy contagion' that it dreamed of.

---

83   M. Antliff, *Inventing Bergson: Cultural Politics and the Parisian Avant-Garde* (Princeton, NJ, 1993).

# Chapter 9
# Conclusion: Literature, Performance, and the Aesthetics of Authority

This book has outlined a series of conceptual 'diagrams' of an ethos of modernity – a stance towards life, truth, and authority – that was styled in the humorous literature, performances and practices that were associated with the cabarets artistiques of fin-de-siècle Montmartre.[1] It has analyzed these experiments in order to contribute to a spatial history of what Foucault called 'arts of living' or 'aesthetics of existence', focusing in particular on the relationship between life, authority, and the experience of truth. In particular, we have seen how, during the nineteenth century, literature and performance came to acquire an important place in experiments with inventing forms of creative truth-telling that deployed novel, 'immanent' kinds of authority to lend weight to their claims. The book has traced the emergence of new forms of 'experiential' authority, where the dynamics and vitality of encounters with the limits of experience offered means for challenging dominant discourses and knowledges concerning the modern city. These literary and artistic experiments with experiential authority can be interpreted as attempts to make productive use of the decline of transcendent structures of authority associated with religion and tradition, and to look instead to the intensities of corporeal experience to ground new kinds of social protest and to create new forms of political and creative community. In the excessive energy and disturbing disjointedness of lived experience, life could find its most dynamic expression; and in the political culture of the Third Republic, life was the ultimate object and objective of government and social action. Through experiments with using humour and parody to undermine dominant truths about the life of the city and emphasize alternative ones, Montmartre artists hoped to resist with authority, voicing new truths concerning the true life of the city.

A key literary figure for these experiments was the 'black cat'. In Chapter 2, I argued that the black cat should be seen as an important figure of the experience of truth in modern French culture. The black cats in Baudelaire, Manet and Poe figured truth as something unrepresentable, something encountered only through absence and via forces beyond the living.

Evoking such unrepresentable truths about the life of the city required literary and artistic performances capable of establishing new forms of authority.

---

1 The notion of 'diagramming' used here draws on Derek McCormack, 'Diagramming Practice and Performance', *Environment and Planning D: Society and Space*, 23 (2005).

In Chapter 3, I addressed the ways in which the perceived 'crisis of authority' after the 1871 civil war triggered experiments with harnessing the authority of life, as expressed in embodied experience, in order to ground new social forms and practices. Whilst such forms of 'experiential authority' were used in reaffirming established biopolitical discourses, they could also be used to contest dominant regimes of power. In Baudelaire's city poetry, for example, we see a way of mobilizing the lived experience of suffering in order to lend authority to a vehement protest against the destruction of experience in the modern city.

In Chapter 4, I started to focus on the literature and performances associated with fin-de-siècle Montmartre, charting the rise of alternative culture in Montmartre in the 1870s. Montmartre, the chapter argued, emerged as a privileged space of liberal governmentality, being allowed a high level of cultural freedom to critique existing social values. From the start, Montmartre alternative culture was associated with aggressive, violent humour. Through humour, Montmartre's writers and artists attempted to reveal and test the limits of the 'life of the city'. Through performance and embodied experience, Montmartre cabarets tested and contested established truths concerning the life of the city.

The cabaret artistique the Chat Noir (Black Cat) was at the heart of these performative interventions in dominant biopolitical urban discourses. Chapter 5 examined the use of humour in the Chat Noir to re-imagine the nature of 'place' and to establish a new affective economy in Montmartre. Through ironic humour and pantomime buffoonery, Montmartre bohemians attempted to negotiate the contradictions of modernity and create an affective engagement with place that articulated both modernity's violence and inequality, but also new kinds of urban life and vitality.

Turning from the affective life of the city to the representational life of the city, Chapter 6 outlined the ethos of 'counter-display' that was discernible in the literary performances of the *Chat Noir* journal and publications. Through false museum catalogues and humorous parodies of colonial literature, these texts dramatized the failure of hegemonic modes of representation within the city of Paris. This ethos of counter-display articulated an alternative experience of life as a form of error: a creative interruption of dominant partitions of visibility and representation.

Moving to the perceptual life of the city, Chapter 7 looked at the spectral, haunted textual and performed landscapes invented at the Chat Noir. Here, Montmartre writers and performers attempted to reveal ghostly forms of life that were invisible to the organic senses, or to derange the senses and unravel the organization of the sensing body. By gesturing towards forms of perception beyond the organic senses, these performances evoked new forms of non-organic life and new kinds of true experience.

As Chapter 8 discussed, the perceived political failure of these forms of cultural resistance, however, led to a renewed emphasis within the Montmartre anarchist movement on direct action and 'propaganda by the deed'. Central to the anarchist movement of the 1890s was an intersection between humour and violence. Violence, like humour, promised to unravel the dominant order of visibilities and invisibilities, creating newly materialized forms of anarchist truth. Humour, by contrast, drew on the authority of intense embodied experiences to lend different forms of legitimacy to the anarchist movement. The movement's failure to reject the dominant biopolitical emphasis on taking life as the final object and objective of government, however, meant that it was unable to pose a long-lasting challenge to hegemonic biopolitical discourses, since both could be reversed and turned back on the anarchists.

The literature, art and performances of the Montmartre cabarets, we have seen, invented a number of new techniques for reimagining the true life of the city. These techniques aimed to contest the ways in which knowledges of the life of the city were being deployed in emerging techniques of government and control. However, they were unable to let go of the notion that life was an ultimate value. Rather than rejecting the biopolitical discourses that centred on life as the final object and objective, they pointed towards other forms of alternative life and vitality. In doing so, however, they participated in the same discourse that they were critiquing, and failed to find grounds for a lasting challenge to emerging structures of government in the Third Republic.

This book has contributed to an improved understanding of the biopolitical experience of truth in the modern city, as well as of the ways in which spatial cultural forms can be used, not only to question existing forms of authority, but also to stylize new forms of authority and political power. It has advanced debates concerning the role of affect, percept and representation in urban biopolitics by contributing to a new analysis of how biopolitical authority works at affective, perceptual and representational registers. It has extended research on the biopolitical constitution of modernity – and in particular the biopolitical structuring of modern experience – by emphasizing the role of creative experiments with the experiential structures of the life of the city in testing and contesting dominant biopolitical discourses and power relations.

In doing so the book aims to open up an agenda for the study of the 'aesthetics of authority' in the modern city. It has argued for a renewed engagement with cultural experiments with reimagining the 'life of the city' that emphasizes how claims about the nature of urban life and vitality acquire or lose authority. Authority is a concept that is associated with top-down, hierarchical, anti-democratic forms of social relationship. Yet in order to resist effectively, it is necessary to resist with authority. There is a need to better understand the ways in which radical or oppositional forms of social and political organization make successful claims to authority. Doing so requires recognizing the plurality of authority, refusing to reduce it to any singular model. In this book I have emphasized that authority can be produced through imaginative literary and performative practices that structure experience in new ways.

This book provides a starting point for a wider project of analyzing the affective, perceptual and representational aesthetics of authority in the modern city, and how these forms of authority generate new experiences of truth through the dynamics of embodied experience. I have drawn upon the example of the

Montmartre creative community, not so much because it offered a successful attempt to engineer a new form of urban life and vitality, but because it played a crucial and under-acknowledged role in the invention of the repertoire of creative practices of resistance in the twentieth century. The rise of the Montmartre cabarets during the 1880s was a pivotal moment in the history of urban cultural politics, anticipating both new forms of popular culture and also the experiments of avant-garde groups such Dada, Surrealism, Situationism, and the cultural politics associated with the revolts of May 1968.[2] Montmartre bohemians used art, I have argued, to build new forms of immanent, 'experiential' authority through which powerful social and political claims for a more egalitarian, more creative and less acquisitive society could be heard. Their cultural experiments invented a new 'embodied imagination', transforming the experience of the city and opposing its rationalization, fabricating a re-enchanted, magical, dynamic counter-modernity.[3] By using novel, often humorous, forms of affective, representational and perceptual experience in order to deploy an alternative experience of truth, they enabled their critique of society (and their utopian vision for a new one), to acquire considerable power and influence.

Montmartre's artistic and anarchist community seem to have had faith in the ability of art to stylize, on its own, a new form of political community. Certainly they did little to back the strength of their aesthetic innovations with the institutional, organizational and practical means through which to implement lasting social change. The anarchist faith in building a new society through individual acts of violence was almost bewilderingly utopian. This perhaps reflects the unresolved tensions between the anti-authoritarian impulse towards ridiculing established political structures and values, and the attempt to create new forms of authority from which to make more effective social demands. Indeed, the move whereby disillusioned anarchists fell back to a poetics of individual violent rebellion indicates the weaknesses they rightly discerned in the cultural politics of Montmartre's avant-garde. The anti-authoritarian impulses that were an essential part of the Montmartre counter-culture were turned against its own attempts to create new forms of authority from which to make radical social and political demands.

Whilst its political ideals remained unfulfilled, and it found its own experiments with urban living subjected to the same anti-authoritarian impulses that it had initiated, the Montmartre counter-culture succeeded in pointing the way to some important emerging structures of authority in the twentieth century. It made a case for the virtues of urban creativity, improvisation and risk-taking that resonated throughout the century, providing inspiration for the radical political movements

---

2   See Anja Kanngieser, *Experimental Politics and the Making of Worlds* (Farnham, 2013), Kristin Ross, *May '68 and Its Afterlives* (Chicago, 2002), Simon Rycroft, *Swinging City: A Cultural Geography of London 1950–1974* (Farnham, 2011).

3   On the embodied imagination, see Leila Dawney, 'Social Imaginaries and Therapeutic Self-Work: The Ethics of the Embodied Imagination', *Sociological Review* 59(3) (2011).

of the Situationists, and hence indirectly influencing the revolts of May '68.[4] It successfully created a myth of egalitarian urban creativity and dynamism that remains powerful in the present day, even where this power is co-opted by narrow economic interests.

What relevance does this historical study have for our understanding of contemporary arts of urban living? In their well-known studies of modern urban experience, Walter Benjamin and later David Harvey turned to the Paris of the Second Empire in order to uncover the spatial dynamics of capital in industrial modernity.[5] It is this Paris of the mid-nineteenth century that has become the iconic image of European modernity. In order to understand the history of the modern *biopolitical* city, however, attention must instead be turned to the Third Republic. It is in the emergence of the Third Republic of France during the last third of the nineteenth century that the key foundations of the modern biopolitical city first started to be laid down.[6] It was during this period that 'society' really emerged as a privileged object of governmental intervention. The notion of 'sécurité sociale' (tempering the revolutionary connotations of the 'sociale' with the respect for law and order evoked by 'sécurité') was invented; the social sciences began to take shape; and the city was reconceived as a social environment, the milieu of a population of living bodies. Even if the physical architecture of the city was now largely in place, the city was emerging in political, popular, and literary discourse in very new ways. And it is this republican city, rather than Haussmann's city, that anticipated many later urban movements.

In recent years, the concept of 'neo-bohemian' space has acquired a degree of prominence in urban regeneration discourses. Richard Florida, for example, celebrates the emergence of a new 'creative class' and emphasizes its members' usefulness for attracting modern, hi-tech and rapid growth industries in search of a labour force.[7] The kinds of environment that this creative class are attracted to, he argues, are bohemian areas with a buzz, vitality and edgy feel, meaning

---

4   Andy Merrifield, 'The Sentimental City: The Lost Urbanism of Pierre Mac Orlan and Guy Debord', *International Journal of Urban and Regional Research* 28(4) (2004).

5   Walter Benjamin, *The Arcades Project*, ed. R. Tiedmann, trans. K. McLaughlin and H. Eiland (Cambridge, MA, 1999), David Harvey, *Paris: Capital of Modernity* (New York, 2003).

6   Paul Rabinow, *French Modern: Norms and Forms of the Social Environment* (1989).

7   Richard Florida, *The Rise of the Creative Class: And How it's Transforming Work, Leisure, Community and Everyday Life* (New York, 2002), Richard Florida, 'Bohemia and Economic Geography', *Journal of Economic Geography* 2(1) (2002), John Hannigan, 'A Neo-Bohemian Rhapsody: Cultural Vibrancy and Controlled Edge as Urban Development Tools in the "New Creative Economy"', in Tim Gibson and Mark Lowes (eds), *Urban Communication: Production, Text, Context* (Lanham, MD, 2006), Phil Hubbard, *City* (2006), Charles Landry, *The Creative City: A Toolkit for Urban Innovators* (London, 2000), Jamie Peck, 'Struggling with the Creative Class', *International Journal of Urban and Regional Research* 29(4) (2005), Andy Pratt, 'Creative Cities: The Cultural Industries and the Creative Class', *Geografiska Annaler B* 90(2) (2008).

that city planners need to be wise to a new 'geography of bohemia' and to plan new 'neo-bohemian' quarters. As Richard Lloyd puts it, 'in neo-bohemia smaller scale cultural offerings and offbeat elements of street level culture are not only important amenities for particular urban consumers, but resources for cultural and new media enterprises'.[8] Indeed, he remarks elsewhere, 'elements of bohemia, surprisingly durable through subsequent generations and still most obviously found in increasingly ubiquitous urban districts, generate dispositions and competencies among adherents that are surprisingly amenable to neoliberal and postindustrial capitalist practices'.[9]

This book has offered a critical perspective on writing lauding the rise of the cultural ecoomy and creative city, showing how bohemian literature and performances in late nineteenth century Montmartre aimed to radically re-imagine the limits of life, rather than simply to repeat existing experiences of urban life and vitality. In the cultural biopolitics of the late nineteenth century we see for the first time a series of attempts to challenge and reconfigure the emerging biopolitical discourses that would define the city of the twentieth century. We see, perhaps, alternative directions that biopolitical discourses and experiences could have taken. We also see the emergence of new forms of experiential authority that would become increasingly important throughout the twentieth century.

The attempts of Montmartre performers to reimagine the life of the city came up against a series of disabling contradictions. Montmartre culture found itself opposing the life of the city in favour of other forms of 'truer', more authentic forms of non-representational life. Rather than rejecting biopolitical discourses entirely, Montmartre bohemians pointed towards newer, better, and more 'modern' forms of life. By doing so, however, their claims to access truer forms of life could always be undermined by the hegemonic biopolitical discourses that framed their experiments as forms of degenerate, unhealthy life. Moreover, their new experiments with deploying the authority of experience merely anticipated the kind of experience economy that would become so dominant later in the following century. If they had instead sought new evaluative criteria that did not take life as a cardinal object and objective of their social and cultural experiments, they might have enjoyed more success in creating a viable alternative to the dominant political forces of the Third Republic. The challenge was, and remains, to invent new, more egalitarian ways of feeling, seeing, and representing the life of the contemporary city.

---

8    Richard Lloyd, 'Neo-Bohemia: Art and Neighborhood Redevelopment in Chicago', *Journal of Urban Affairs* 24(5) (2002), p. 157.

9    Richard Lloyd, 'Beyond the Protestant Ethic: Culture, Subjectivity and Instrumental Labor', in David Chalcraft et al. (eds), *Max Weber Matters: Interweaving Past and Present* (Farnham, 2008), p. 206.

# References

A'Kempis, 'Coup d'état du 2 Novembre 1882', *Le Chat Noir*, 28 October 1882.
——, 'L'assaut De Montmartre', *Le Chat Noir*, 1 April 1882.
——, 'Voyages De Découvertes', *Le Chat Noir*, 14 January 1882.
——, 'Voyages De Découvertes', *Le Chat Noir*, 21 January 1882.
——, 'Voyages De Découvertes (Suite)', *Le Chat Noir*, 4 February 1882.
Abrahamsson, Christian and Abrahamsson, Sebastian, 'In Conversation With the Body Conveniently Known as Stelarc', *Cultural Geographies* 14 (2007), pp. 293–398.
Adorno, Theodor, *Negative Dialectics*, trans. E. Ashton (London: Routledge, 1973).
——, *In Search of Wagner* (London: Verso, 2005).
Adorno, Theodor and Max Horkheimer, *Dialectic of Enlightenment*, trans. J. Cumming (London: Verso, 1997).
Agulhon, Maurice, 'Paris: A Traversal from East to West', in Pierre Nora (ed.), *Realms of Memory: The Construction of the French Past, Volume III, Symbols* (New York: Columbia University Press, 1998).
Aisenberg, Andrew, *Contagion: Disease, Government, and the 'Social Question' in Nineteenth-Century France* (Stanford, CA: Stanford University Press, 1999).
Amin, Ash and Nigel Thrift, *Cities: Reimagining the Urban* (Cambridge: Polity, 2002).
Anderson, Ben, 'Affect and Biopower: Towards a Politics of Life', *Transactions of the Institute of British Geographers* 37 (2012), pp. 28–43.
Anderson, Robert David, *France 1870–1914: Politics and Society* (London: Routledge and Kegan Paul, 1977).
Anon., 'La Colère du Mont Aventin', *Le Mont-Aventin: Organe Quotidien de la Fédération Républicaine* 1871.
Anon., 'Echos de l'escalier', *L'Anti-Concierge: Organe Officiel de la Défense des Locataires* 1 (1881).
Anon., 'Le Chat Noir', *Le Chat Noir*, 8 April 1881.
Anon., 'Marseillaise des Locataires', *L'Anti-Concierge: Organe Officiel de la Défense des Locataires* 1 (1881).
Anon., 'L'Etat Naturel et la Part du Proletaire dans la Civilisation', (Archives de la Prefecture de la Police, n.d.).
Ansell-Pearson, Keith, 'Free Spirits and Free Thinkers: Nietzsche and Guyau on the Future of Morality', in Jeffrey Metzger (ed.), *Nietzsche, Nihilism and the Philosophy of the Future* (London: Continuum, 2009), pp. 102–24.
Antliff, M., *Inventing Bergson: Cultural Politics and the Parisian Avant-Garde* (Princeton, NJ: Princeton University Press, 1993).

Appignanesi, Lisa, *The Cabaret* (London: Methuen, 1984).

Aquarone, Stanislas, *The Life and Works of Emile Littré 1801–1881* (Leyden: A. W. Sythoff, 1958).

Arendt, Hannah, 'What Is Authority?', *Between Past and Future: Six Exercises in Political Thought* (New York: Viking Press, 1961), pp. 91–142.

——, 'On Violence', *Crises of the Republic* (Harmondsworth: Penguin, 1973), pp. 103–84.

Artus, Maurice, 'Essai de bibliographie de la presse Montmartoise: journaux et canards', *Bulletin de la societé d'histoire et d'archaeologie des IXè et XVIIIè Arrondissements: Le Vieux Montmartre* 49–50 (1906).

Aubery, P., 'The Anarchism of the Literati of the Symbolist Period', *The French Review* 42 (1968), pp. 39–47.

Auriol, Georges, *Le Chat Noir – Guide* (Paris: Le Chat Noir, 1887).

——, 'Rodolphe Salis et les deux "Chat Noir"', *Mercure de France* 9 (1926), pp. 315–32.

Bailes, Sara Jane, *Performance Theatre and the Poetics of Failure: Forced Entertainment, Goat Island, Elevator Repair Service* (London: Routledge, 2011).

Bain, Alexander, *The Senses and the Intellect* (London: J. W. Parker & Son, 1855).

——, *The Emotions and the Will* (London: J. W. Parker and Son, 1859).

Bakhtin, Mikhail, *Rabelais and His World*, trans. Hélène Iswolsky (Bloomington, IN: Indiana University Press, 1984).

Bakunin, Mikhail, *God and the State* (New York: Dover Publications, 1970).

Barbey D'Aurevilly, Jules Amédée, 'Maurice Rollinat', *Le Constitutitionel*, 1 June 1882.

Barnes, David, *The Great Stink of Paris and the Nineteenth-Century Struggle against Filth and Germs* (Baltimore: Johns Hopkins University Press, 2006).

Barrucand, Victor, 'Le Rire De Ravachol', *L'Endehors* 64 (1892).

Bashford, Alison, 'Global Biopolitics and the History of World Health', *History of the Human Sciences* 19 (2006), pp. 67–88.

Bataille, Georges, *Manet*, trans. Austryn Wainhouse and James Emmons (London and Basingstoke: Macmillan, 1983).

Bataille, Georges and Annette Michelson, 'Un-Knowing: Laughter and Tears', *October* 36 (1986), pp. 89–102.

Baudelaire, Charles, 'On the Essence of Laughter', in Jonathan Mayne (ed.), *The Painter of Modern Life, and Other Essays* (London: Phaidon, 1964).

——, 'The Painter of Modern Life', in Jonathan Mayne (ed.), *The Painter of Modern Life, and Other Essays* (London: Phaidon, 1964).

——, *The Flowers of Evil*, trans. James McGowan (Oxford: Oxford University Press, 1993).

——, *Paris Spleen and La Fanfarlo*, trans. Raymond MacKenzie (Indianapolis: Hackett, 2008).

Bauman, Zygmunt, *Liquid Modernity* (Cambridge: Polity, 2000).

——, *The Art of Life* (Cambridge: Polity, 2008).

Beiser, Frederick, *The Romantic Imperative: The Concept of Early German Romanticism* (Cambridge, MA: Harvard University Press, 2003).

——, *Hegel* (London and New York: Routledge, 2005).

Beizer, Janet, *Ventriloquized Bodies: Narratives of Hysteria in Nineteenth-Century France* (Ithaca: Cornell University Press, 1994).

Benjamin, Walter, 'Central Park', *New German Critique* 34 (1985), pp. 32–58.

——, *Charles Baudelaire: A Lyric Poet in the Era of High Capitalism* (London: Verso, 1997).

——, *The Origin of German Tragic Drama*, trans. J. Osborne (London: Verso, 1998).

——, *The Arcades Project*, trans. K. McLaughlin and H. Eiland (Cambridge, MA: Harvard University Press, 1999).

——, 'On Some Motifs in Baudelaire', in Hannah Arendt (ed.), *Illuminations* (London: Pimlico, 1999), pp. 152–96.

——, 'The Storyteller', in Hannah Arendt (ed.), *Illuminations* (London: Pimlico, 1999).

——, 'Theses on the Philosophy of History', in Hannah Arendt (ed.), *Illuminations* (London: Pimlico, 1999), pp. 245–55.

——, 'The Work of Art in the Age of Its Technological Reproducibility', in Howard Eiland and Michael Jennings (eds) *Walter Benjamin: Selected Writings, Volume Three, 1935–1938* (Cambridge, MA: Harvard University Press, 2002).

——, 'The Paris of the Second Empire in Baudelaire', in Howard Eiland and Michael Jennings (eds) *Walter Benjamin: Selected Writings, Volume Four, 1938–1940* (Cambridge, MA: Harvard University Press, 2006), pp. 3–92.

——, 'Protocols of Drug Experiments', *On Hashish* (Cambridge, MA: Belknap, 2006).

Bennett, Tony, *Birth of the Museum: History, Theory, Politics* (London and New York: Routledge, 1995).

Bercy, Anne de and Armand Ziwès, *À Montmartre ... Le Soir. Cabarets et Chansonniers D'hier* (Paris: Ed. Grasset, 1951).

Bergson, Henri, *Matter and Memory*, trans. W. Palmer and Nancy Paul (London: Macmillan, 1911).

——, *Comedy* (New York: Doubleday, 1956).

Berman, Marshall, *All That Is Solid Melts into Air: The Experience of Modernity* (London: Verso, 1983).

Bernard, Claude, *Introduction À L'Étude de la Médecine Expérimentale* (Paris, 1865).

Bernheimer, Charles, *Figures of Ill Repute: Representing Prostitution in Nineteenth-Century France* (Cambridge, MA: Harvard University Press, 1989).

——, *Decadent Subjects: The Idea of Decadence in Art, Literature, Philosophy, and Culture of the Fin-De-Siècle in Europe* (Baltimore: Johns Hopkins University Press, 2001).

Bhambra, Gurminder, *Rethinking Modernity: Postcolonialism and the Sociological Imagination* (Basingstoke: Palgrave Macmillan, 2007).

Bihl, Laurent, 'L'«Armée du Chahut»: Les Deux Vachalcades de 1896 et 1897', *Sociétés & Représentations* 27 (2001).

Billig, Michael, 'Humour and Embarrassment', *Theory, Culture & Society* 18 (2001), pp. 23–43.

——, *Laughter and Ridicule: Towards a Social Critique of Humour* (London: Sage, 2005).

Bissell, David, Maria Hynes and Scott Sharpe, 'Unveiling Seductions Beyond Societies of Control: Affect, Security, and Humour in Spaces of Aeromobility', *Environment and Planning D: Society & Space* 30 (2012), pp. 694–710.

Blanchot, Maurice, *The Infinite Conversation*, trans. Susan Hanson (Minneapolis: University of Minnesota Press, 1993).

Blencowe, Claire, 'Destroying Duration: The Critical Situation of Bergsonism in Benjamin's Philosophy of Modern Experience', *Theory, Culture & Society* 25 (2008), pp. 137–56.

——, *Biopolitical Experience: Foucault, Power and Positive Critique* (Basingstoke: Palgrave, 2012).

——, 'Biopolitical Authority, Objectivity and the Groundwork of Modern Citizenship', *Journal of Political Power* 6 (2013), pp. 9–28.

Blencowe, Claire, Julian Brigstocke and Leila Dawney (eds), 'Authority and Experience', *Special Issue of Journal of Political Power* 6 (2013).

Bogad, L. M., 'Carnivals against Capital: Radical Clowning and the Global Justice Movement', *Social Identities* 16 (2010), pp. 537–57.

Boime, Albert, *Art and the French Commune: Imagining Paris after War and Revolution* (Princeton, NJ: Princeton University Press, 1995).

Bonnefoy, Yves, 'Baudelaire's *Les Fleurs Du Mal*', in John Naughton (ed.), *The Act and the Place of Poetry: Selected Essays* (Chicago: University of Chicago Press, 1989), pp. 44–9.

Bonnett, Alastair, 'Situationism, Geography, and Poststructuralism', *Environment and Planning D: Society and Space* 7 (1989), pp. 131–46.

——, 'Art, Ideology and Everyday Space: Subversive Tendencies from Dada to Postmodernism', *Environment and Planning D: Society and Space* 10 (1992), pp. 69–86.

——, 'The Nostalgias of Situationist Subversion', *Theory, Culture & Society* 23 (2006), pp. 23–48.

——, 'The Dilemmas of Radical Nostalgia in British Pyschogeography', *Theory, Culture & Society* 26 (2009), pp. 47–72.

Breton, Andre, *Anthology of Black Humor* (San Francisco: City Lights Books, 1997).

Brigstocke, Julian, 'Defiant Laughter: Humour and the Aesthetics of Place in Late 19th Century Montmartre', *Cultural Geographies* 19 (2012), pp. 217–35.

——, 'Artistic Parrhesia and the Genealogy of Ethics in Foucault and Benjamin', *Theory, Culture & Society* 30 (2013), pp. 57–78.

——, 'Immanent Authority and the Performance of Community in Late Nineteenth Century Montmartre', *Journal of Political Power* 6 (2013), pp. 107–26.

Bryant, Chad, 'The Language of Resistance? Czech Jokes and Joke-Telling under Nazi Occupation, 1943–45', *Journal of Contemporary History* 41 (2006), pp. 133–51.

Buci-Glucksmann, Christine, 'Catastrophic Utopia: The Feminine as Allegory of the Modern', *Representations* 14 (1986), pp. 220–29.

Buck-Morss, Susan, 'The Flâneur, the Sandwichman and the Whore: The Politics of Loitering', *New German Critique* 39 (1986), pp. 99–140.

——, *The Dialectics of Seeing: Walter Benjamin and the Arcades Project* (Cambridge, MA: MIT Press, 1989).

Buffon, Georges, *Histoire Naturelle, Générale et Particulière* (Paris: De l'Imprimerie Royale, 1769).

Buisson, Sylvie and Christian Parisot, *Paris Montmartre: Les Artistes and Les Lieux, 1860–1920* (Paris: Terrail, 1996).

Burckhardt, Jacob, *The Civilisation of the Period of the Renaissance in Italy*, trans. S. Middlemore (London: Kegan Paul & Co., 1878).

Bürger, Peter, *Theory of the Avant-Garde*, trans. Michael Shaw (Manchester: Manchester University Press, 1984).

Bury, J. P. T., *Gambetta's Final Years: The Era of Difficulties, 1877–1882* (London: Longman, 1982).

Butler, Judith, *Excitable Speech: A Politics of the Performative* (New York; London: Routledge, 1997).

Cahen, Fabrice, 'Medicine, Statistics, and the Encounter of Abortion and "Depopulation" in France (1870–1920)', *History of the Family* 14 (2008), pp. 19–35.

Călinescu, Matei, *Five Faces of Modernity: Modernism, Avant-Garde, Decadence, Kitsch, Postmodernism* (Durham, NC: Duke University Press, 1987).

Callon, Michel, Pierre Lascoumes and Yannick Barthe, *Acting in an Uncertain World: An Essay on Technical Democracy* (Cambridge, MA: MIT, 2009).

Candida Smith, Richard, *Mallarmé's Children: Symbolism and the Renewal of Experience* (Berkeley, CA: University of California Press, 1999).

Canguilhem, Georges, *On the Normal and the Pathological*, trans. Carolyn Fawcett (Dordrecht, Holland: D. Reidel, 1978).

——, 'The Living and Its Milieu', *Grey Room* 3 (2001), pp. 6–31.

Caradec, François and Alain Weill, *Le Café-Concert* (Paris: Atelier Hachette/ Massin, 1980).

Carroll, Joseph, *Evolution and Literary Theory* (Columbia: University of Missouri Press, 1995).

Castells, M., 'Cities and Revolution: The Commune of Paris, 1871', *The City and the Grassroots: A Cross-Cultural Theory of Urban Social Movements* (Berkeley: University of California Press, 1983).

Cate, Phillip Dennis, 'The Spirit of Montmartre', in Phillip Dennis Cate and Mary Shaw (eds) *The Spirit of Montmartre: Cabarets, Humor and the Avant-Garde, 1875–1905* (New Brunswick, NJ: Rutgers University Press, 1996), pp. 1–94.

——, 'The Social Menagerie of Toulouse-Lautrec's Montmartre', in Richard Thomson, Phillip Dennis Cate and Mary Weaver Chapin (eds) *Toulouse-Lautrec and Montmartre* (Washington: National Gallery of Art 2005), pp. 26–45.

Cate, Phillip Dennis and Mary Shaw (eds), *The Spirit of Montmartre: Cabarets, Humor, and the Avant-Garde, 1875–1905* (New Brunswick, NJ: Rutgers University Press, 1996).

Caygill, Howard, *A Kant Dictionary* (Oxford: Blackwell, 1995).

——, *Walter Benjamin: The Colour of Experience* (London: Routledge, 1998).

Chadwick, Kay, 'Education in Secular France: (Re)Defining *Laïcité*', *Modern & Contemporary France* 5 (1997), pp. 47–59.

Chanouard, 'Il Faut Lutter', *Le Chat Noir*, 8 April 1882.

Chapin, Mary Weaver, 'The Chat Noir & the Cabarets', in Richard Thomson, Phillip Dennis Cate and Mary Weaver Chapin (eds) *Toulouse-Lautrec and Montmartre* (Washington: National Gallery of Art 2005), pp. 89–107.

Charlton, Donald Geoffrey, *Positivist Thought in France During the Second Empire, 1852–1870* (Oxford: Clarendon Press, 1959).

Chevalier, Louis, *Montmartre du Plaisir et du Crime* (Paris: Éditions Robert Laffont, 1995).

Chevalier, Louis, *Labouring Classes and Dangerous Classes in Paris During the First Half of the Nineteenth Century*, trans. Frank Jellinek (London: Routledge and Kegan Paul, 1973).

Choay, Francoise, *The Modern City: Planning in the 19th Century*, trans. George R. Collins and Marguerite Hugo (London: Studio Vista, 1969).

Claretie, Jules, *Histoire de la Révolution de 1870–71* (Paris: Aux Bureaux du Journal *L'Éclipse*, 1872).

Clarke, David and Marcus Doel, 'Engineering Space and Time: Moving Pictures and Motionless Trips', *Journal of Historical Geography* 31 (2005), pp. 41–60.

Clarke, Timothy J., *The Painting of Modern Life: Paris in the Art of Manet and His Followers* (New York: Knopf, 1985).

Cole, Sarah, 'Dynamite Violence and Literary Culture', *Modernism/Modernity* 16 (2009), pp. 301–28.

Colebrook, Claire, *Irony* (London: Routledge, 2004).

Comte, Auguste, *A General View of Positivism* (Routledge, 1908).

Copleston, Frederick, *A History of Philosophy Volume 9: Maine De Biran to Sartre* (New York: Image Books, 1977).

Corbin, Alain, *Women for Hire: Prostitution and Sexuality in France after 1850* (Cambridge, MA: Harvard University Press, 1990).

Coulson, Shea, 'Funnier Than Unhappiness: Adorno and the Art of Laughter', *New German Critique* 34 (2007), pp. 141–63.

Cowan, Bainard, 'Walter Benjamin's Theory of Allegory', *New German Critique* (1983).

Crary, Jonathan, *Techniques of the Observer: On Vision and Modernity in the Nineteenth Century* (Cambridge, MA: MIT Press, 1990).

——, *Suspensions of Perception: Attention, Spectacle, and Modern Culture* (Cambridge, MA: MIT Press, 1999).

Cresswell, Tim, *Place: A Short Introduction* (Oxford: Blackwell, 2004).

——, 'Laughter and the Tramp', *The Tramp in America* (London: Reaktion, 2013), pp. 130–70.

Crimp, Douglas, 'On the Museum's Ruins', *October* 13 (1980), pp. 41–57.

Critchley, Simon, *On Humour* (London: Routledge, 2001).

——, *Infinitely Demanding: Ethics of Commitment, Politics of Resistance* (London: Verso, 2007).

da Costa, Beatriz and Philip, Kavita (eds), *Tactical Biopolitics: Art, Activism, and Technoscience* (Cambridge, MA: MIT Press, 2008)

Darwin, Charles, *The Expression of the Emotions in Man and Animals* (London: Murray, 1872).

Daston, Lorraine and Peter Galison, *Objectivity* (New York: Zone, 2007).

Datta, Venita, 'A Bohemian Festival: La Fête De La Vache Enragée', *Journal of Contemporary History* 28 (1993), pp. 195–213.

Davies, Peter, *The Extreme Right in France, 1789 to the Present: From De Maistre to Le Pen* (London: Routledge, 2002).

Dawney, Leila, 'Social Imaginaries and Therapeutic Self-Work: The Ethics of the Embodied Imagination', *Sociological Review* 59 (2011), pp. 535–52.

——, 'The Figure of Authority: The Affective Biopolitics of the Mother and the Dying Man', *Journal of Political Power* 6 (2013), pp. 29–47.

de Marc, Félix, *Légitimité et la Révolution: Étude sur le Principe D'Autorité* (Paris: Maurice Tardieu, 1882).

Deak, F., *Symbolist Theater: The Formation of an Avant-Garde* (Baltimore: Johns Hopkins University Press, 1993).

Debord, Guy, *Society of the Spectacle* (Detroit: Black & Red, 1983).

Debré, Patrice, *Louis Pasteur* (Baltimore: Johns Hopkins University Press, 1998).

Deleuze, Gilles and Félix Guattari, *A Thousand Plateaus: Capitalism and Schizophrenia, Volume Two*, trans. Brian Massumi (London and New York: Continuum, 2004).

Dennis, Richard, *Cities in Modernity: Representations and Productions of Metropolitan Space, 1840–1930* (Cambridge: Cambridge University Press, 2008).

Derrida, Jacques, 'The Animal That Therefore I Am (More to Follow)', *Critical Inquiry* 28 (2002), pp. 369–418.

Deutsche, Rosalyn, 'Boys Town', *Environment and Planning D: Society & Space* 9 (1991), pp. 5–30.

Dews, Peter, 'The Return of the Subject in Late Foucault', *Radical Philosophy* 51 (1989), pp. 37–41.

Dewsbury, J.-D. 'Affective Habit Ecologies: Material Dispositions and Immanent Inhabitations', in *Performance Research: A Journal of the Performing Arts*, 17 (2012), pp. 74–82.

Dewsbury, J.-D., 'Avant-Garde/Avant-Garde Geographies', in Rob Kitchen & Nigel Thrift (eds) *International Encyclopedia of Human Geography* (Amsterdam: Elsevier, 2009), pp. 252–6.

Dienstag, Joshua, 'Nietzsche's Dionysian Pessimism', *The American Political Science Review* 95 (2001), pp. 923–37.

Dilts, Andrew, 'From "Entrepreneur of the Self" to "Care of the Self": Neoliberal Governmentality and Foucault's Ethics', *Foucault Studies* 12 (2011), pp. 130–46.

Dixon, Deborah, 'Creating the Semi-living: On Politics, Aesthetics and the More-than-human', *Transactions of the Institute of British Geographers* 34 (2009), pp. 411–25.

Dodds, Klaus and Philip Kirby, 'It's Not a Laughing Matter: Critical Geopolitics, Humour and Unlaughter', *Geopolitics* 18 (2013), pp. 45–59.

Driver, Felix, *Geography Militant: Cultures of Exploration and Empire* (Oxford: Blackwell, 2001).

du Seigneur, Marcus, 'Vitruve et Gambrinus et Le Chat Noir', *La Construction Moderne* (1885), pp. 517–19.

Duncan, C. and A. Walloch, 'The Universal Survey Museum', *Art History* 3 (1980), pp. 447–69.

Duncan, James and Derek Gregory, *Writes of Passage: Reading Travel Writing* (London: Routledge, 1999).

Eagleton, Terry, *The Ideology of the Aesthetic* (Oxford: Blackwell, 1990).

——, *The Idea of Culture* (Oxford: Blackwell, 2000).

Eisenstadt, S. N., 'Multiple Modernities', *Daedalus* 129 (2000), pp. 1–29.

Elden, S., *Mapping the Present: Heidegger, Foucault and the Project of a Spatial History* (London: Continuum, 2001).

Ellingson, Terry, *The Myth of the Noble Savage* (Berkeley, CA: University of California Press, 2001).

Ellis, Jack, *The Physician-Legislators of France: Medicine and Politics in the Early Third Republic, 1870–1914* (Cambridge: Cambridge University Press, 1990).

Emery, Elizabeth and L. Morowitz, *Consuming the Past: The Medieval Revival in Fin-De-Siècle France* (Aldershot: Ashgate, 2003).

Epstein, Kate and Kurt Iveson, 'Locking Down the City (Well, Not Quite): Apec 2007 and Urban Citizenship in Sydney', *Australian Geographer* 40 (2009), pp. 271–95.

Esslin, Martin, *The Theatre of the Absurd* (Garden City, NY: Doubleday, 1961).

Evans, Martha Noel, *Fits and Starts: A Genealogy of Hysteria in Modern France* (Ithaca, NY: Cornell University Press, 1991).

Faudemay, Alain, 'L'humour et L'anarchisme: Quelques Indices d'une Convergence Possible', *Revue D'Histoire Littéraire de la France* 99 (1999), pp. 467–84.

Florida, Richard, 'Bohemia and Economic Geography', *Journal of Economic Geography* 2 (2002), pp. 55–71.

——, *The Rise of the Creative Class: And How it's Transforming Work, Leisure, Community and Everyday Life* (New York: Basic Books, 2002).

——, *Cities and the Creative Class* (New York and Abingdon: Routledge, 2005).

Forgioni, N., '"The Shadow Only": Shadow and Silhouette in Late Nineteenth Century Paris', *The Art Bulletin* 81 (1999), pp. 490–512.

Foucault, Michel, *The Order of Things: An Archaeology of the Human Sciences* (London: Tavistock, 1970).

——, 'The Politics of Health in the Eighteenth Century', in Paul Rabinow (ed.), *The Foucault Reader* (Harmondsworth: Penguin, 1984), pp. 273–89.

——, *The Care of the Self: The History of Sexuality, Volume Three*, trans. R. Hurley (London: Penguin, 1990).

——, *The Use of Pleasure: The History of Sexuality, Volume Two*, trans. R. Hurley (London: Penguin, 1992).

——, *The Will to Knowledge: The History of Sexuality, Volume One* (London: Penguin, 1998).

——, 'Life: Experience and Science', in Paul Rabinow (ed.), *Aesthetics, Method, and Epistemology: Essential Works of Foucault 1954–1984, Volume Two* (New York: The New Press, 2000), pp. 465–78.

——, 'On the Genealogy of Ethics: An Overview of a Work in Progress', in Paul Rabinow (ed.), *Ethics, Subjectivity and Truth: Essential Works of Foucault 1954–1984, Volume One* (London: Penguin, 2000), pp. 253–80.

——, 'What Is Enlightenment?', in Paul Rabinow (ed.), *Ethics, Subjectivity and Truth: The Essential Works of Foucault 1954–1984, Volume One* (London: Penguin, 2000), pp. 303–19.

——, *Fearless Speech* (Los Angeles: Semiotext, 2001).

——, *The Archaeology of Knowledge*, trans. A.M. Sheridan Smith (London: Routledge, 2002).

——, 'Interview with Michel Foucault', in James Faubion (ed.), *Power. Essential Works of Foucault, 1954–1984, Volume Three* (London: Penguin, 2002), pp. 239–97.

——, *Society Must Be Defended*, trans. D. Macy (London: Penguin, 2004).

——, *The Hermeneutics of the Subject: Lectures at the Collège de France 1981–1982* trans. Graham Burchell (Basingstoke: Palgrave Macmillan, 2005).

——, *Security, Territory, Population: Lectures at the Collège de France, 1977–1978*, trans. Graham Burchell (Basingstoke: Palgrave Macmillan, 2007).

——, *The Birth of Biopolitics: Lectures at the Collège de France, 1978–1979*, trans. Graham Burchell (Basingstoke and New York: Palgrave Macmillan, 2008).

——, *The Government of Self and Others* (Basingstoke: Palgrave Macmillan, 2010).

——, *The Government of Self and Others: Lectures at the Collège de France, 1982–1983* (Basingstoke: Palgrave Macmillan, 2010).

——, *The Courage of Truth: The Government of Self and Others, Volume Two. Lectures at the Collège De France, 1983–1984* (Basingstoke: Palgrave Macmillan, 2011).

Fouillée, Alfred, *Critique des Systèmes de Morale Contemporains* (Paris: Baillère, 1883).

Fouquier, Henri, 'Review of Ubu Roi', *Le Figaro*, 11 December 1896.

Fraser, Benjamin, 'Toward a Philosophy of the Urban: Henri Lefebvre's Uncomfortable Application of Bergsonism', *Environment and Planning D: Society and Space* 26 (2008), pp. 338–58.

Fried, Michael, *Manet's Modernism, or, the Face of Painting in the 1860s* (Chicago: University of Chicago Press, 1996).

Frisby, David, *Fragments of Modernity: Theories of Modernity in the Work of Simmel, Kracauer, and Benjamin* (Cambridge, MA: MIT Press, 1988).

Gaden, Élodie, 'Mutations Des Pratiques Poétiques: 1878–1914, Maurice Rollinat, Marie Krysinska, Valentine De Saint-Point' (Mémoire de Master 2 de recherche, Université Stendhal, 2009).

Gandy, Matthew, 'The Paris Sewers and the Rationalization of Urban Space', *Transactions of the Institute of British Geographers* 24 (1999), pp. 23–44.

——, 'Zones of Indistinction: Bio-Political Contestations in the Urban Arena', *Cultural Geographies* 13 (2006), pp. 497–516.

Garber, Frederick (ed.), *Romantic Irony* (Budapest: Akadémiai Kiadó, 1988).

Gargano, James, 'The Question of Poe's Narrators', *College English* 25 (1963), pp. 177–81.

Gautier, Théophile, *Tableaux du Siège, Paris, 1870–1871* (Bibliothèque-Charpentier: Paris, 1872).

——, *Charles Baudelaire: His Life*, trans. Guy Thorne (New York: Brentano, 1915).

Gilbert, David, Matless, David, and Short, Brian (eds) *Geographies of British Modernity* (Oxford: Blackwell, 2003).

Gendron, Bernard, *Between Montmartre and the Mudd Club: Popular Music and the Avant-Garde* (Chicago: University of Chicago Press, 2002).

Georgette, 'Courrier Musical', *Le Chat Noir* 45, 46, 47, 48 (1882).

Giddens, A., *Modernity and Self-Identity: Self and Society in the Late Modern Age* (Cambridge: Polity, 1991).

Gieryn, Thomas, 'A Space for Place in Sociology', *Annual Review of Sociology* 26 (2000), pp. 463–96.

Gluck, Mary, 'The Flâneur and the Aesthetic Appropriation of Urban Culture in Mid-19th-Century Paris', *Theory, Culture & Society* 20 (2003), pp. 53–80.

——, *Popular Bohemia: Modernism and Urban Culture in Nineteenth-Century Paris* (Cambridge, MA: Harvard University Press, 2005).

Goffman, Erving, 'Embarrassment and Social Organization', *Interaction Ritual: Essays on Face-to-Face Behaviour* (New York: Doubleday, 1967).

Good, Robert, 'The Philosophy and Social Thought of Alfred Fouillée' (PhD thesis, McGill University, 1993).

Gordon, Rae Beth, 'Le Caf Conc' et l'hystérie', *Romantisme* 64 (1989), pp. 53–67.

——, 'From Charcot to Charlot: Unconscious Imitation and Spectatorship in French Cabaret and Early Cinema', *Critical Inquiry* 27 (2001).

——, *Why the French Love Jerry Lewis: From Cabaret to Early Cinema* (Stanford, CA: Stanford University Press, 2001).

——, *Dances with Darwin, 1875–1910: Vernacular Modernity in France* (Farnham: Ashgate, 2009).

Goudeau, Émile, *Dix Ans se Bohème* (Paris: Librairie Illustrée, 1888).

Grand-Carteret, John, *Raphael et Gambrinus, ou L'Art dans la Brasserie* (Paris: L. Westhausser, 1886).

Greenblatt, Stephen, *Renaissance Self-Fashioning: From More to Shakespeare* (Chicago and London: University of Chicago Press, 1980).

Greenhalgh, Paul, *Ephemeral Vistas: The Expositions Universelles, Great Exhibitions and World's Fairs, 1851–1939* (Manchester: Manchester University Press, 1988).

Gregory, Derek, 'Scripting Egypt: Orientalism and the Cultures of Travel', in James Duncan and Derek Gregory (eds) *Writes of Passage: Reading Travel Writing* (London and New York: Routledge, 1999), pp. 114–50.

Guggenheim, Michael, 'Laboratizing and De-Laboratizing the World', *History of the Human Sciences* 25 (2012), pp. 99–118.

Gullickson, Gay, *Unruly Women of Paris: Images of the Commune* (Ithaca, NY: Cornell University Press, 1996).

Guyau, Jean-Marie, *Education and Heredity: A Study in Sociology*, trans. W. J. Greenstreet (Walter Scott, 1891).

——, *A Sketch of Morality Independent of Obligation or Sanction*, trans. G. Kapteyn (London: Watts & Co., 1898).

——, *The Non-Religion of the Future: A Sociological Study* (New York: Shocken, 1962).

Halberstam, Judith, *The Queer Art of Failure* (Durham, NC: Duke University Press, 2011).

Halévy, Daniel, *Pays Parisiens: Portrait de la France* (Paris: Éditions Émile-Paul Frères, 1929).

——, *La Fin Des Notables* (Paris: Grasset, 1930).

Hallward, Peter, 'The Limits of Individuation, or How to Distinguish Deleuze and Foucault', *Angelaki* 5 (2000), pp. 93–111.

Halperin, J., *Félix Fénéon: Aesthete & Anarchist in Fin-De-Siècle Paris* (New Haven and London: Yale University Press, 1988).

Hamilton, George Heard, *Manet and His Critics* (New Haven: Yale University Press, 1954).

Hammett, Daniel, 'Zapiro and Zuma: A Symptom of an Emerging Constitutional Crisis in South Africa?', *Political Geography* 29 (2010), pp. 88–96.

——, 'Resistance, Power and Geopolitics in Zimbabwe', *Area* 43 (2011), pp. 202–10.

Han, Béatrice, *Foucault's Critical Project: Between the Transcendental and the Historical*, trans. E. Pyle (Stanford, CA: Stanford University Press, 2002).

Hannah, Matthew, *Governmentality and the Mastery of Territory in Nineteenth-Century America* (Cambridge: Cambridge University Press, 2000).

——, 'Biopower, Life and Left Politics', *Antipode* 43 (2011), pp. 1034–55.

Hannigan, John, 'A Neo-Bohemian Rhapsody: Cultural Vibrancy and Controlled Edge as Urban Development Tools in the "New Creative Economy"', in Tim Gibson and Mark Lowes (eds) *Urban Communication: Production, Text, Context* (Lanham, MD: Rowman and Littlefield, 2006).

Hansen, Miriam, 'Benjamin's Aura', *Critical Enquiry* 34 (2008), pp. 336–75.

Haraucourt, Edmond, 'La Cité Morte', *Le Chat Noir*, 27 May 1882.

Harvey, David, 'Monument and Myth', *Annals of the Association of American Geographers* 69 (1979), pp. 362–81.

——, *The Condition of Postmodernity: An Enquiry into the Origins of Cultural Change* (Oxford: Basil Blackwell, 1989).

——, 'From Space to Place and Back Again', in *Justice, Nature, and the Geography of Difference* (Oxford: Blackwell, 1996), pp. 291–328.

——, *Paris: Capital of Modernity* (New York: Routledge, 2003).

Harvey, David Allen, 'Beyond Enlightenment: Occultism, Politics, and Culture in France from the Old Regime to the Fin De Siècle', *The Historian* 65 (2003), pp. 665–94.

Hazareesingh, Sudhir, *Intellectual Founders of the Republic: Five Studies in Nineteenth-Century French Political Thought* (Oxford: Oxford University Press, 2001).

Hegel, Georg Wilhelm Friedrich, *Phenomenology of Spirit*, trans. A. V. Miller (Oxford: Clarendon Press, 1977).

Henderson, J., *The First Avant-Garde 1887–1894: Sources of the Modern French Theatre* (London: George G. Harrap & Co., 1971).

Hetherington, Kevin, *Capitalism's Eye: Cultural Spaces of the Commodity* (Abingdon: Routledge, 2007).

Hetherington, Kevin and Monica Degen (eds), 'Spatial Hauntings,' Special edition of *Space and Culture*, 1(11/12), (2001).

Hetherington, Kevin *The Badlands of Modernity: Heterotopia and Social Ordering* (London: Routledge, 2002).

Hewitt, Nicholas, 'Shifting Cultural Centres in Twentieth-Century Paris', in M. Sheringham (ed.), *Parisian Fields* (London: Reaktion, 1996), pp. 30–45.

——, 'From "Lieu De Plaisir" to "Lieu De Mémoire": Montmartre and Parisian Cultural Topography', *French Studies* LIV (2000), pp. 453–67.

Hobbes, Thomas, 'Human Nature', in J. C. A. Gaskin (ed.), *Human Nature and De Corpore Politico* (Oxford: Oxford University Press, 1994).

Hoolsema, Daniel, 'The Echo of an Impossible Future in "the Literary Absolute"', *MLN* 119 (2004), pp. 845–68.

Hooper-Greenhill, Eilean, *Museums and the Shaping of Knowledge* (New York: Routledge, 1992).

Hubbard, Phil, *City* (London: Routledge, 2006).

Huebner, Steven, *French Opera at the Fin De Siècle: Wagnerism, Nationalism, and Style* (Oxford: Oxford University Press, 1999).

Hutton, J., *Neo-Impressionism and the Search for Solid Ground: Art, Science and Anarchism in Fin-De-Siecle France* (Baton Rouge: Louisiana State University Press, 1994).

Huxley, Margo, 'Spatial Rationalities: Order, Environment, Evolution and Government', *Social & Cultural Geography* 7 (2006), pp. 771–87.

Hyman, Erin Williams, 'Theatrical Terror: Attentats and Symbolist Spectacle', *The Comparatist* 29 (2005), pp. 101–22.

Hynes, Maria and Scott Sharpe, 'Yea-Saying Laughter', *Parallax* 16 (2010), pp. 44–54.

Hynes, Maria, Scott Sharpe and Bob Fagan, 'Laughing with the Yes Men: The Politics of Affirmation', *Continuum* 21 (2007), pp. 107–21.

Jackson, Jeffrey, 'Artistic Community and Urban Development in 1920s Montmartre', *French Politics, Culture & Society* 24 (2006), pp. 1–25.

Jakobson, Roman and Claude Lévi-Strauss, '"Les Chats" De Charles Baudelaire', *L'Homme* 2 (1962), pp. 5–21.

James, Ian, 'Naming the Nothing: Nancy and Blanchot on Community', *Culture, Theory and Critique* 51 (2010), pp. 171–87.

Jay, Martin, *Downcast Eyes: The Denigration of Vision in Twentieth-Century French Thought* (Berkeley, CA: University of California Press, 1993).

——, 'Scopic Regimes of Modernity', *Force Fields: Between Intellectual History and Cultural Critique* (New York: Routledge, 1993), pp. 114–33.

——, *Songs of Experience: Modern American and European Variations on a Universal Theme* (Berkeley, CA: University of California Press, 2005).

Jeanne, P., *Les Théâtres D'Ombres à Montmartre de 1887 à 1923* (Paris: Les Editions des Presses Modernes, 1937).

Jellinek, Frank, *The Paris Commune of 1871* (London: Gollancz, 1971).

Jennings, Michael, 'On the Banks of a New Lethe: Commodification and Experience in Benjamin's Baudelaire Book', in Kevin McLaughlin and Philip Rosen (eds) *Benjamin Now: Critical Encounters with the Arcades Project* (Durham, NC: Duke University Press, 2003).

Jeunet, Jean-Pierre (dir.), *Le Fabuleux Destin D'Amélie Poulain*, produced by Jean-Marc Deschamps and Claudie Ossard (UGC & Miramax Films, 2001), 123 minutes.

Johnson, Peter, 'Foucault's Spatial Combat', *Environment and Planning D: Society & Space* 26 (2008), pp. 611–26.

Jonas, Raymond, 'Sacred Tourism and Secular Pilgrimage: Montmartre and the Basilica of Sacré Coeur', in Gabriel Weisberg (ed.), *Montmartre and the Making of Mass Culture* (New Brunswick, NJ: Rutgers University Press, 2001), pp. 94–119.

Jones, Louisa, *Sad Clowns and Pale Pierrots: Literature and the Popular Comic Arts in 19th-Century France* (Lexington, KY: French Forum, 1984).

Joyce, Patrick, *The Rule of Freedom: Liberalism and the Modern City* (London: Verso, 2003).

Jullian, Philippe, *Montmartre*, trans. Anne Carter (Oxford: Phaidon, 1977).

Justice-Malloy, Rhona, 'Charcot and the Theatre of Hysteria', *The Journal of Popular Culture* 28 (1995), pp. 133–8.

Kanngieser, Anja, *Experimental Politics and the Making of Worlds* (Farnham: Ashgate, 2013).

Kant, Immanuel, 'What Is Enlightenment?', in L. Beck (ed.), *Selections* (New York: Macmillan, 1988), pp. 462–7.

——, *The Critique of Judgment*, trans. J. H. Bernard (New York: Prometheus, 2000).

Kehily, Mary Jane and Anoop Nayak, '"Lads and Laughter": Humour and the Production of Heterosexual Hierarchies', *Gender and Education* 9 (1997), pp. 69–88.

Kenny, N., 'Je Cherche Fortune: Identity, Counterculture, and Profit in Fin-De-Siècle Montmartre', *Urban History Review* 32 (2004).

Kete, Kathleen, *The Beast in the Boudoir: Petkeeping in Nineteenth-Century Paris* (Berkeley, CA: University of California Press, 1994).

Kirwan, Samuel 'On the "Inoperative Community" and Social Authority: A Nancean Response to the Politics of Loss', *Journal of Political Power* 6(1) (2013), pp. 69–86.

Knutson, Harold, '*An Enemy of the People:* Ibsen's Reluctant Comedy', *Comparative Drama* 27 (1993), pp. 159–75.

Koven, Seth, *Slumming: Sexual and Social Politics in Victorian London* (Princeton: Princeton University Press, 2004).

Kraftl, Peter, 'Liveability and Urban Architectures: Mol(ecul)ar Biopower and the "Becoming Lively" of Sustainable Communities', *Environment and Planning D: Society and Space,* 32 (2014).

Kropotkin, Peter, *Ethics: Origin and Development*, trans. Louis S. Friedland and Joseph R. Piroshnikoff (New York: Lincoln MacVeagh, 1924).

——, 'Anarchist Morality', in Roger Baldwin (ed.), *Kropotkin's Revolutionary Pamphlets* (New York: Dover Publications, 1970).

——, 'Modern Science and Anarchism', in Roger Baldwin (ed.), *Kropotkin's Revolutionary Pamphlets* (New York: Dover Publications, 1970), pp. 145–94.

——, *Mutual Aid: A Factor of Evolution* (London: Freedom Press, 1987).

Krysinska, Marie, 'Symphonie en Gris', in André Velter (ed.), *Les Poètes du Chat Noir* (Paris: Gallimard, 1996), pp. 233–4.

Kuipers, Giselinde, *Good Humor, Bad Taste: A Sociology of the Joke* (Berlin and New York: Mouton de Gruyter, 2006).

Kunzle, David, 'The Voices of Silence: Willette, Steinlen and the Introduction of the Silent Strip in the *Chat Noir*, with a German Coda', in Robin Varnum and Christina Gibbons (eds) *The Language of Comics: Word and Image* (Jackson: University Press of Mississippi, 2001), pp. 3–18.

'La Chanson Traquée. Fermeture du cabaret du "Pierrot Noir"', *Le Matin*, 18 April 1897.

Lacoue-Labarthe, Philippe, *Heidegger, Art and Politics: The Fiction of the Political*, trans. Chris Turner (Oxford: Basil Blackwell, 1990).

Lacoue-Labarthe, Philippe and Jean-Luc Nancy, *The Literary Absolute: The Theory of Literature in German Romanticism*, trans. Philip Barnard and Cheryl Lester (Albany: State University of New York Press, 1988).

Landry, Charles, *The Creative City: A Toolkit for Urban Innovators* (London: Earthscan, 2000).

Lanson, Gustave, *Histoire de la Literature Francaise* (Paris: Librairie Hachette, 1912).

Larousse, Pierre, 'Chat', *Grand Dictionnaire Universel du XIX Siècle, Tome Troisième* (Paris: Administration du Grand Dictionnaire Universel, 1867), pp. 1064–8.

Latour, Bruno, *The Pasteurization of France*, trans. Alan Sheridan and John Law (Cambridge, MA: Harvard University Press, 1988).

Lawlor, Leonard, *The Implications of Immanence: Towards a New Concept of Life* (New York: Fordham University Press, 2006).

Lay, Howard, '"Beau Geste!" (on the Readability of Terrorism)', *Yale French Studies* 101 (2001), pp. 79–100.

Lazzarato, Maurizio, 'Neoliberalism in Action: Inequality, Insecurity and the Reconstitution of the Social', *Theory, Culture & Society* 26 (2009), pp. 109–33.

Le Feuvre, Lisa (ed.), *Failure: Documents of Contemporary Art* (Cambridge, MA: MIT Press, 2010).

Lefebvre, Henri, *La Proclamation De La Commune: 26 Mars 1871* (Paris: Gallimard, 1965).

——, *The Production of Space*, trans. Donald Nicholson-Smith (Oxford: Blackwell, 1991).

——, *Critique of Everyday Life, Volume Two: Foundations for a Sociology of the Everyday*, trans. John Moore (London and New York: Verso, 2002).

Legg, Stephen, *Spaces of Colonialism: Delhi's Urban Governmentalities* (Oxford: Blackwell, 2007).

Lehardy, Jacques, 'Lettres D'un Explorateur', *Le Chat Noir*, 18 March 1882.

——, 'Montmartre', *Le Chat Noir*, 14 January 1882.

Lehning, J., *To Be a Citizen: The Political Culture of the Early French Third Republic* (Ithaca, NY: Cornell University Press, 2001).

Lemaître, Jules, 'Le Chat-Noir', *Les Gaîtés Du Chat Noir* (Paris: Paul Ollendorff, n.d.), pp. 5–6.

Lévêque, Jean-Jacques, *Les Années Impressionnistes* (Courbevoie: ACR, 1990).

Levin, David Michael (ed.), *Modernity and the Hegemony of Vision* (Berkeley, CA: University of California Press, 1993).

Ley, David, 'Artists, Aestheticisation and the Field of Gentrification', *Urban Studies* 40 (2003), pp. 2527–44.

Lisbonne, Maxime, 'La Taverne du Bagne', *Gazette du Bagne* 1 (1885).

Littre, Émile, *Application de la Philosophie Positive au Gouvernement des Societes et en Particulier a la Crise Actuelle* (Paris: Librairie Philosophique de Ladrange, 1850).

——, 'La Centralisation', *Journal des Débats*, 7 and 11 October 1862.

Lloyd, Richard, 'Neo-Bohemia: Art and Neighborhood Redevelopment in Chicago', *Journal of Urban Affairs* 24 (2002), pp. 517–32.

——, *Neo-Bohemia: Art and Commerce in the Postindustrial City* (London: Routledge, 2005).

——, 'Beyond the Protestant Ethic: Culture, Subjectivity and Instrumental Labor', in David Chalcraft, Fanon Howell, Marisol Lopez Memendez and Hector Vera (eds) *Max Weber Matters: Interweaving Past and Present* (Farnham: Ashgate, 2008), pp. 205–20.

Lloyd, Rosemary (ed.), *Selected Letters of Charles Baudelaire: The Conquest of Solitude* (Chicago: University of Chicago Press, 1986).

Locke, Robert R., *French Legitimists and the Politics of Moral Order in the Early Third Republic* (Princeton: Princeton University Press, 1974).

Lockyer, Sharon and Michael Pickering, 'You Must Be Joking: The Sociological Critique of Humour and Comic Media', *Sociology Compass* 2 (2005), pp. 808–20.

—— (eds), *Beyond a Joke: The Limits of Humour* (Basingstoke: Palgrave Macmillan, 2005).

Lombroso, César, 'L'Anarchie et ses Héros', *La Revue des revues*, 15 February 1894.

Lovell, Terry, 'Resisting with Authority: Historical Specificity, Agency and the Performative Self', *Theory, Culture & Society* 20 (2003), pp. 1–17.

Loyer, François, *Paris Nineteenth Century: Architecture and Urbanism* (New York: Abbeville Press, 1988).

Lucas, John, 'From Naturalism to Symbolism', *Renaissance and Modern Studies* 21 (1977), pp. 124–39.

Luhrmann, Baz (dir.), *Moulin Rouge!*, produced by Fred Baron, Martin Brown and Baz Luhrmann (20th Century Fox, 2001), 128 minutes.

Luxon, Nancy, *Crisis of Authority: Politics, Trust, and Truth-Telling in Freud and Foucault* (Cambridge: Cambridge University Press, 2013).

——, 'Truthfulness, Risk, and Trust in the Late Lectures of Michel Foucault', *Inquiry* 57 (2004), pp. 464–89.

Macmillan, Alexandre, 'Michel Foucault's Techniques of the Self and the Christian Politics of Obediance', *Theory, Culture & Society* 28 (2011), pp. 3–25.

Macnab, Maurice, 'Ballade Des Accents Circonflexes', *Poèmes Mobiles* (Paris, 1890).

Macpherson, Hannah, '"I Don't Know Why They Call It the Lake District. They Might as Well Call It the Rock District!" The Workings of Humour and Laughter in Research with Members of Visually Impaired Walking Groups', *Environment and Planning D: Society and Space* 26 (2008), pp. 1080–95.

Mainardi, Patricia, *The End of the Salon: Art and the State in the Early Third Republic* (Cambridge: Cambridge University Press, 1993).

Mansell Jones, P., 'Poe and Baudelaire: The "Affinity"', *The Modern Language Review* 40 (1945), pp. 279–83.

Markus, Gyorgy, 'Walter Benjamin Or: The Commodity as Phantasmagoria', *New German Critique* 83 (2001), pp. 3–42.

Martin, Lauren L., 'Bombs, Bodies, and Biopolitics: Securitizing the Subject at the Airport Security Checkpoint', *Social & Cultural Geography* 11 (2010), pp. 17–34.

Marx, Karl, *Economic and Philosophical Manuscripts of 1844* (Moscow: Foreign Languages Printing House, 1961).

Massumi, Brian, *Parables for the Virtual: Movement, Affect, Sensation* (Durham, NC: Duke University Press, 2002).

Maupassant, Guy de, *Chroniques Inédites* (Paris: Maurice Gonon, 1979).

McClellan, Andrew, *Inventing the Louvre: Art, Politics, and the Origins of the Modern Museum in Eighteenth-Century Paris* (Cambridge: Cambridge University Press, 1994).

McCormack, Derek, 'An Event of Geographical Ethics in Spaces of Affect', *Transactions of the Institute of British Geographers*, 28 (2003), pp. 488–507.

—— 'Diagramming Practice and Performance', *Environment and Planning D: Society and Space*, 23 (2005), pp. 119–47.

McHugh, Kevin and Ann Fletchall, 'Festival and the Laughter of Being', *Space and Culture* 15 (2012), pp. 381–94.

McKinley, Alexander, *Illegitimate Children of the Enlightenment: Anarchists and the French Revolution, 1880–1914* (New York: P. Lang, 2008).

McNay, Lois, 'The Foucauldian Body and the Exclusion of Experience', *Hypatia* 6 (1991), pp. 125–39.

Menon, Elizabeth, 'Potty-Talk in Parisian Plays: Henry Somm's *La Berline De L'Émigré* and Alfred Jarry's *Ubu Roi*', *Art Journal* 52 (1993), pp. 59–64.

——, 'Images of Pleasure and Vice: Women of the Fringe', in Gabriel Weisberg (ed.), *Montmartre and the Making of Mass Culture* (New Brunswick, NJ: University of Rutgers Press, 2001), pp. 37–71.

Merrifield, Andy, 'The Sentimental City: The Lost Urbanism of Pierre Mac Orlan and Guy Debord', *International Journal of Urban and Regional Research* 28 (2004), pp. 930–40.

Merriman, John, *The Dynamite Club: How a Bombing in Fin-De-Siècle Paris Ignited the Age of Modern Terror* (London: JR Books, 2009).

Millner, Naomi, 'Routing the Camp: Experiential Authority in a Politics of Irregular Migration', *Journal of Political Power*, 6(1) (2013), pp. 87–105.

Mitchell, Timothy, *Colonising Egypt* (Berkeley: University of California Press, 1988).

——, 'Orientalism and the Exhibitionary Order', in Nicholas Dirks (ed.), *Colonialism and Culture* (Ann Arbor: University of Michigan Press, 1992), pp. 289–318.

——, *Rule of Experts: Egypt, Techno-Politics, Modernity* (Berkeley, CA: University of California Press, 2002).

Moréas, Jean, 'Le Symbolisme', *Le Figaro, Supplément Littéraire*, 18 September 1886.

Morreall, John, *The Philosophy of Laughter and Humor* (Albany, N.Y.: State University of New York Press, 1987).

Munholland, John Kim, 'Republican Order and Republican Tolerance in Fin-De-Siècle France: Montmartre as Delinquent Community', in Gabriel Weisberg (ed.), *Montmartre and the Making of Mass Culture* (New Brunswick, NJ: Rutgers University Press, 2001).

Murard, Lion and Patrick Zylberman, *L'Hygiène dans la République: La Santé Publique en France, ou, L'Utopie Contrariée, 1870–1918* (Paris: Fayard, 1996).

Murphy, Richard, *Theorizing the Avant-Garde: Modernism, Expressionism, and the Problem of Postmodernity* (Cambridge: Cambridge University Press, 1999).

Nally, David, '"That Coming Storm": The Irish Poor Law, Colonial Biopolitics, and the Great Famine', *Annals of the Association of American Geographers* 98 (2008), pp. 714–41.

Nancy, Jean-Luc, *The Inoperative Community*, trans. Peter Connor, Lisa Garbus, Michael Holland and Simona Sawhney (Minneapolis: University of Minnesota Press, 1991).

Nechayev, Sergey, 'The Revolutionary Catechism', available at http://www. marxists.org/subject/anarchism/nechayev/catechism.htm, accessed: 6 April 2014. (1989).

Nerval, Gerard de, *Promenades et Souvenirs. Lettres à Jenny Pandora* (Paris: Garnier-Flammarion, 1972).

Newman, Saul, 'Anarchism and the Politics of Ressentiment', *Theory & Event* 4 (2000).

Nietzsche, Friedrich, *On the Genealogy of Morals*, trans. W. Kaufmann and R. J. Hollingdale (New York: Random House, 1969).

——, *The Birth of Tragedy*, trans. Shaun Whiteside (London: Penguin, 1993).Noël, Benoît, *L'Absinthe: Une Fée Franco-Suisse* (Yens sur Morges: Cabédita 2001).

Noorani, Tehseen, 'Service User Involvement, Authority and the 'Expert-by-Experience' in Mental Health', *Journal of Political Power* 6 (2013), pp. 49–68.

Noorani, Tehseen, Claire Blencowe and Julian Brigstocke (eds), *Problems of Participation: Reflections on Authority, Democracy and the Struggle for Common Life* (Lewes: ARN Press, 2013).

Nora, Pierre (ed.), *Realms of Memory: The Construction of the French Past* (New York: Columbia University Press, 1998).

Nye, Robert, *Crime, Madness & Politics in Modern France: The Medical Concept of National Decline* (Princeton: Princeton University Press, 1984).

O'Farrell, Clare, *Foucault: Historian or Philosopher?* (Basingstoke: Macmillan, 1980).

O'Gorman, Róisín and Margaret Werry, 'On Failure (on Pedagogy): Editorial Introduction', *Performance Research* 17 (2012), pp. 1–8.

Oberthur, Mariel, *Le Cabaret du Chat Noir à Montmartre* (Geneva: Slatkine, 2007).

Oerlemans, Onno, *Romanticism and the Materiality of Nature* (Toronto: University of Toronto Press, 2002).

Offen, Karen, 'Depopulation, Nationalism, and Feminism in Fin-De-Siècle France', *The American Historical Review* 89 (1984), pp. 648–76.

Ogborn, Miles, *Spaces of Modernity: London's Geographies, 1680-1780* (New York: Guildford Press, 1998).

Oksala, Johanna, 'Violence and the Biopolitics of Modernity', *Foucault Studies* 10 (2010), pp. 23–43.

Orlay, Gustave, *L'autorité & la liberté opposées au despotisme du nombre* (Paris: J. Féchoz, 1874).

Orru, Marcho, 'The Ethics of Anomie: Jean Marie Guyau and Émile Durkheim', *The British Journal of Sociology* 34 (1983), pp. 499–518.

Osborne, Peter, *The Politics of Time: Modernity and the Avant-Garde* (London: Verso, 1995).

Osborne, Thomas, 'Security and Vitality: Drains, Liberalism and Power in the Nineteenth Century', in Andrew Barry, Thomas Osborne and Nikolas Rose (eds) *Foucault and Political Reason: Liberalism, Neo-Liberalism, and Rationalities of Government* (Chicago: University of Chicago Press, 1996).

——, *Aspects of Enlightenment: Social Theory and the Ethics of Truth* (London: UCL Press, 1998).

——, 'Against 'Creativity': A Philistine Rant', *Economy and Society* 32 (2003), pp. 507–25.

Osborne, Thomas and Nikolas Rose, 'Governing Cities: Notes on the Spatialisation of Virtue', *Environment and Planning D: Society and Space* 17 (1999), pp. 737–60.

——, 'Spatial Phenomenotechnics: Making Space with Charles Booth and Patrick Geddes', *Environment and Planning D: Society and Space* 22 (2004), pp. 209–28.

Osgood, Samuel M., *French Royalism Since 1870* (The Hague: Martinus Nijhoff, 1970).

Péchu, Cécile, 'Entre Résistance et Contestation: La Genèse du Squat comme Mode D'Action', *Université de Lausanne Travaux de Science Politique* 24 (2006).

Peck, Jamie, 'Struggling with the Creative Class', *International Journal of Urban and Regional Research* 29 (2005).

Pick, Daniel, *Faces of Degeneration: A European Disorder, C. 1848–C. 1918* (Cambridge: Cambridge University Press, 1989).

Pile, Steve, *Real Cities: Modernity, Space and the Phantasmagorias of City Life* (London: Sage, 2005).

Pillet, Elisabeth, 'Cafés-Concerts et Cabarets', *Romantisme* 75 (1992), pp. 43–50.

Pinder, David, 'Subverting Cartography: The Situationists and Maps of the City', *Environment and Planning A* 28 (1996), pp. 405–27.

——, '"Old Paris Is No More": Geographies of Spectacle and Anti-Spectacle', *Antipode* 32 (2000), pp. 357–86.

——, *Visions of the City: Utopianism, Power and Politics in Twentieth-Century Urbanism* (Edinburgh: Edinburgh University Press, 2005).

Pinkney, D., *Napoleon III and the Rebuilding of Paris* (Princeton, NJ: Princeton University Press, 1958).

Poe, Edgar Allan, 'The Black Cat', *The Raven and Other Poems and Tales* (Boston, New York: Bullfinch Press, 2001), pp. 26–38.

Poggioli, R., *Theory of the Avant-Garde*, trans. G. Fitzgerald (Boston: Harvard University Press, 1968).

Powell, Chris and George Paton (eds), *Humour in Society: Resistance and Control* (Basingstoke: Macmillan, 1988).

Pratt, Andy, 'Creative Cities: The Cultural Industries and the Creative Class', *Geografiska Annaler B* 90 (2008), pp. 107–17.

Pratt, Mary Louise, *Imperial Eyes: Travel Writing and Transculturation* (London: Routledge, 1992).

Prestwich, Patricia, *Drink and the Politics of Social Reform: Antialcoholism in France since 1870* (Palo Alto: Society for the Promotion of Science and Scholarship, 1988).

Purcell, Darren, Melissa Scott Brown and Mahmut Gokmen, 'Achmed the Dead Terrorist and Humor in Popular Geopolitics', *GeoJournal* 75 (2010), pp. 373–85.

Rabinow, Paul, *French Modern: Norms and Forms of the Social Environment* (Cambridge, MA: MIT Press, 1989).
——, 'Introduction: A Vital Rationalist', in François Delaporte (ed.), *A Vital Rationalist: Selected Writings from Georges Canguilhem* (New York: Zone Books, 1994).
——, 'Foucault's Untimely Struggle: Toward a Form of Spirituality', *Theory, Culture & Society* 26 (2009), pp. 25–44.
Rancière, Jacques, 'The Aesthetic Revolution and Its Outcomes: Emplotments of Autonomy and Heteronomy', *New Left Review* 14 (2002).
——, *The Politics of Aesthetics: The Distribution of the Sensible*, trans. G. Rockhill (London: Continuum, 2004).
Rearick, Charles, *Pleasures of the Belle Epoque: Entertainment & Festivity in Turn of the Century France* (New Haven: Yale University Press, 1985).
Reff, Theodore, *Manet: Olympia* (London: Allen Lane, 1976).
Rémond, R., *The Right Wing in France. From 1815 to De Gaulle*, trans. James Michael Laux (Philadelphia: University of Philadelphia Press, 1966).
Ribot, Théodule, *La Psychologie Anglaise Contemporaine* (Librairie Philosophique de Ladrange: Paris, 1870).
Richard, Noel, *A L'Aube du Symbolisme: Hydropathes, Fumistes et Décadents* (Paris: Librairie Nizet, 1961).
Richardson, Joanna, *The Courtesans: The Demi-Monde in Nineteenth-Century France* (London: Weidenfeld & Nicolson, 1967).
Ridanpää, Juha, 'Geopolitics of Humour: The Muhammed Cartoon Crisis and the Kaltio Comic Strip Episode in Finland', *Geopolitics* 14 (2009), pp. 729–49.
Rimbaud, Arthur, *Illuminations and Other Prose Poems* (New York: New Directions, 1957).
Robb, Graham, '"Les Chats" de Baudelaire: Une Nouvelle Lecture', *Revue D'Histoire Littéraire de la France* 6 (1985), pp. 1002–10.
Rochard, Eugène, 'L'Avenir de L'Hygiène', *Revue Scientifique* 24 (1887), pp. 387–95.
Rodrigues, S. B. and D. L. Collinson, '"Having Fun"? Humour as Resistance in Brazil', *Organisation Studies* 16 (1995), pp. 739–68.
Rollinat, Maurice, 'L'Amante Macabre', *Les Névroses* (Paris: G. Charpentier, 1885), pp. 255–8.
——, 'La Peur', *Les Névroses* (Paris: G. Charpentier, 1885), pp. 249–54.
——, 'Le Néant', *L'Abîme* (Paris: G. Charpentier, 1886).
Ronell, Avital, *Loser Sons: Politics and Authority* (Champaign, IL: University of Illinois Press, 2012).
Rose-Redwood, R., 'Governmentality, Geography, and the Geo-Coded World', *Progress in Human Geography* 30 (2006), pp. 469–86.
Rose, Nikolas, 'Medicine, History and the Present', in Colin Jones and Roy Porter (eds) *Reassessing Foucault: Power, Medicine, and the Body* (London: Routledge, 1994), pp. 48–72.
——, *Powers of Freedom: Reframing Political Thought* (Cambridge: Cambridge University Press, 1999).

Roslak, Robyn, *Neo-Impressionism and Anarchism in Fin-De-Siècle France: Painting, Politics and Landscape* (Aldershot: Ashgate, 2007).

Ross, Kristin, *The Emergence of Social Space: Rimbaud and the Paris Commune* (Minneapolis: University of Minnesota Press, 1988).

——, *May '68 and Its Afterlives* (Chicago: University of Chicago Press, 2002).

Rounding, Virginia, *Grandes Horizontales: The Lives and Legends of Four Nineteenth-Century Courtesans* (London: Bloomsbury, 2003).

Rousseau, Jean-Jacques, *Essay on the Origin of Languages and Writings Related to Music* (Hanover, NH: University Press of New England, 1998).

Routledge, Paul, 'Sensuous Solidarities: Emotion, Politics and Performance in the Clandestine Insurgent Rebel Clown Army', *Antipode* 44 (2012), pp. 428–52.

Rycroft, Simon, *Swinging City: A Cultural Geography of London 1950–1974* (Farnham: Ashgate, 2011).

Salis, Rodolphe, 'Poster for the Montmartre Municipal Election', (Paris, 1884).

Salzani, Carlo, 'The City as Crime Scene: Walter Benjamin and the Traces of the Detective', *New German Critique* 34 (2007), pp. 165–87.

Sanders, Teela, 'Controllable Laughter: Managing Sex Work through Humour', *Sociology* 38 (2004), pp. 273–91.

Sartre, Jean-Paul, *Baudelaire* (New York: New Directions, 1967).

Schlegel, Friedrich von, *Philosophical Fragments*, trans. Peter Firchow (Minneapolis: University of Minnesota Press, 1991).

Schwartz, Vanessa, *Spectacular Realities: Early Mass Culture in Fin-De-Siècle Paris* (Berkeley: University of California Press, 1998).

Scott Haine, W., *The World of the Paris Cafe: Sociability among the French Working Class, 1789–1914* (Baltimore: The Johns Hopkins University Press, 1996).

Scott, John, *Republican Ideas and the Liberal Tradition in France, 1870–1914* (New York: Columbia University Press, 1951).

Segel, Harold, 'Fin De Siècle Cabaret', *Performing Arts Journal* 2 (1977), pp. 41–57.

——, *Turn-of-the-Century Cabaret: Paris, Barcelona, Berlin, Munich, Vienna, Cracow, Moscow, St. Petersburg, Zürich* (New York: Columbia University Press, 1987).

Seigel, Jerrold, *Bohemian Paris: Culture, Politics, and the Boundaries of Bourgeois Life, 1830–1930* (New York: Viking, 1986).

Sennett, Richard, *Authority* (London: Secker and Warburg, 1980).

Shapiro, Ann-Louise, *Housing the Poor of Paris* (Madison: University of Wisconsin Press, 1985).

Sharpe, Scott, Maria Hynes and Robert Fagan, 'Beat Me, Whip Me, Spank Me, Just Make It Right Again: Beyond the Didactic Masochism of Global Resistance', *Fibreculture* 6 (2005).

Shattuck, R., *The Banquet Years: The Origins of the Avant-Garde in France 1885–1918* (New York: Vintage Books, 1968).

Shaw, Mary, 'All or Nothing? The Literature of Montmartre', in Phillip Dennis Cate and Mary Shaw (eds) *The Spirit of Montmartre: Cabarets, Humor and the Avant-Garde, 1875–1905* (New Brunswick, NJ: Rutgers University Press, 1996), pp. 111–58.

Shepard, Benjamin, *Play, Creativity and Social Movements: If I Can't Dance, It's Not My Revolution* (London: Routledge, 2011).

Sheringham, Michael, *Everyday Life: Theories and Practices from Surrealism to the Present* (Oxford: Oxford University Press, 2006).

Sherman, Daniel, *Worthy Monuments: Art Museums and the Politics of Culture in Nineteenth-Century France* (Cambridge, MA: Harvard University Press, 1989).

Shryock, Richard 'Becoming Political: Symbolist Literature and the Third Republic', *Nineteenth Century French Studies* 33 (1958), pp. 385–98.

Simmel, Georg, 'Sociology of the Senses: Visual Interaction', in Robert Park and Ernest Burgess (eds) *Introduction to the Science of Sociology* (Chicago: University of Chicago Press, 1924).

——, 'The Concept and Tragedy of Culture', in David Frisby and Mike Featherstone (eds) *Simmel on Culture: Selected Writings* (London: Sage, 1997), pp. 55–74.

Simonsen, Kirsten, 'Bodies, Sensations, Space and Time: The Contribution from Henri Lefebvre', *Geografiska Annaler B* 87 (2005), pp. 1–14.

Simpson, Martin, 'The Death of Henri V: Legitimists without the Bourbons', *French History* 15 (2001), pp. 378–99.

Simpson, Paul, 'Ecologies of Experience: Materiality, Sociality, and the Embodied Experience of (Street) Performing', *Environment and Planning A*, 45 (2013), pp. 180–96.

Singer, Peter, *A Darwinian Left: Politics, Evolution and Cooperation* (London: Weidenfeld & Nicolson, 1999).

Sloterdijk, Peter, *Critique of Cynical Reason* (Minneapolis, MA: University of Minnesota Press, 1987).

Somm, Henri, *La Berline de L'Emigré, ou Jamais Trop Tard Pour Bien Faire* (Paris: Le Chat Noir, 1885).

Sonn, Richard, *Anarchism and Cultural Politics in Fin-De-Siècle France* (Lincoln, NE: University of Nebraska Press, 1989).

——, 'Marginality and Transgression: Anarchy's Subversive Allure', in Gabriel Weisberg (ed.) *Montmartre and the Making of Mass Culture* (New Brunswick, NJ: University of Rutgers Press, 2001), pp. 120–41.

Soppelsa, Peter, *The Fragility of Modernity: Infrastructure and Everyday Life in Paris, 1870–1914* (PhD thesis, University of Michigan, 2009).

Sorel, Georges, *Reflections on Violence*, trans. Jeremy Jennings (Cambridge: Cambridge University Press, 1999).

Spencer, Herbert, 'The Physiology of Laughter', *Essays: Scientific, Political and Speculative* (New York: D. Appleton, 1864).

Spiegel, Gabrielle, 'The Cult of Saint Denis and Capetian Kingship', *Journal of Medieval History* 1 (1975), pp. 43–69.

Springer, Anne-Marie, 'Terrorism and Anarchy: Late 19th Century Images of a Political Phenomenon in France', *Art Journal* 38 (1979), pp. 261–6.

Synnott, Anthony, 'The Eye and I: A Sociology of Sight', *International Journal of Politics, Culture, and Society* 5 (1992), pp. 617–36.

t'Hart, Marjolein and Dennis Bos (eds), *Humour and Social Protest* (Cambridge: Cambridge University Press, 2007).

Tabarin, 'Au Théâtre du Chat Noir', *Le Magazine Français Illustrée*, 25 March 1891.

Tanke, Joseph, *Foucault's Philosophy of Art: A Genealogy of Modernity* (London: Continuum, 2009).

Tester, Keith (ed.), *The Flâneur* (London: Routledge, 1994).

Thompson, Richard, 'Introducing Montmartre', in Richard Thomson, Phillip Dennis Cate and Mary Weaver Chapin (eds) *Toulouse-Lautrec and Montmartre* (Washington: National Gallery of Art, 2005).

Thomson, Richard, Phillip Dennis Cate and Mary Weaver Chapin (eds), *Toulouse-Lautrec and Montmartre* (Washington: National Gallery of Art, 2005).

Thrift, Nigel, 'Lifeworld Inc.: And What to do About it', *Environment and Planning D: Society and Space*, 29 (2011), pp. 5–26.

——, 'Steps to an Ecology of Place', in Doreen Massey, John Allen and Philip Sarre (eds) *Human Geography Today* (Cambridge: Polity Press, 1999), pp. 295–322.

——, *Non-Representational Theory: Space, Politics, Affect* (London: Routledge, 2007).

——, 'Overcome by Space: Reworking Foucault', in Jeremy Crampton and Stuart Elden (eds) *Space, Knowledge and Power: Foucault and Geography* (Aldershot: Ashgate, 2007), pp. 53–8.

Tombs, Robert, *The Paris Commune, 1871* (London: Longman, 1999).

Tönnies, Ferdinand, *Community and Civil Society*, trans. Jose Harris and Margaret Hollis (Cambridge: Cambridge University Press, 2001).

Trahair, Lisa, 'The Comedy of Philosophy: Bataille, Hegel and Derrida', *Angelaki* 6 (2001): 155–69.

Valbel, Horace, *Les Chansonniers et les Cabarets Artistiques* (Paris: E. Dentu, 1895).

Valéry, Paul, 'The Triumph of Manet', in *Degas, Manet, Morisot*, trans. D. Paul (Princeton, NJ: Princeton University Press, 1960).

van Reigen, Willem, 'Breathing the Aura – the Holy, the Sober Breath', *Theory, Culture & Society* 18 (2001), pp. 31–50.

Van Zanten, D., *Building Paris: Architectural Institutions and the Transformation of the French Capital 1830–1870* (Cambridge: Cambridge University Press, 1994).

Vasudevan, Alexander, 'Symptomatic Acts, Experimental Embodiments: Theatres of Scientific Protest in Interwar Germany', *Environment and Planning A* 39 (2007), pp. 1812–37.

Veber, Pierre, 'Les Cabarets Artistiques et la Chanson', *La Revue D'Art Dramatique*, 15 December 1889.

Wagner, Richard, *The Art Work of the Future* (Whitefish, MT: Kessinger, 2004).

Weaver, Simon, 'The "Other" Laughs Back: Humour and Resistance in Anti-Racist Comedy', *Sociology* 44 (2010), pp. 31–48.

——, *The Rhetoric of Racist Humour: US, UK and Global Race Joking* (Farnham: Ashgate, 2011).

Weber, Max, *The Theory of Social and Economic Organization.*, trans. A. M. Henderson and Talcott Parsons (London: Collier-Macmillan, 1964).

Weir, David, *Anarchy & Culture: The Aesthetic Politics of Modernism* (Amherst: University of Massachusetts Press, 1997).

Weisberg, Gabriel, 'Montmartre's Lure: An Impact on Mass Culture', in Gabriel Weisberg (ed.), *Montmartre and the Making of Mass Culture* (New Brunswick, NJ: Rutgers University Press, 2001).

——, (ed.), *Montmartre and the Making of Mass Culture* (New Brunswick, NJ: University of Rutgers Press, 2001).

Wernick, Andrew, *Auguste Comte and the Religion of Humanity: The Post-Theistic Program of French Social Theory* (Cambridge: Cambridge University Press, 2001).

Whiting, Steven Moore, *Satie the Bohemian: From Cabaret to Concert Hall* (Oxford: Oxford University Press, 1999).

Whitmore, Janet, 'Absurdist Humor in Bohemia', in Gabriel Weisberg (ed.), *Montmartre and the Making of Mass Culture* (New Brunswick, NJ: University of Rutgers Press, 2001), pp. 205–22.

Williams, E., 'Signs of Anarchy: Aesthetics, Politics, and the Symbolist Critic at the *Mercure de France*, 1890–95', *French Forum* 29 (2004), pp. 45–68.

Williams Hyman, Erin, 'Theatrical Terror: Attentats and Symbolist Spectacle', *The Comparatist* 29 (2005), pp. 101–22.

Willis, Paul, *Learning to Labour: How Working Class Kids Get Working Class Jobs* (Aldershot: Ashgate, 1977).

Wilson, Colette, *Paris and the Commune, 1871–78: The Politics of Forgetting* (Manchester: Manchester University Press, 2007).

Wilson, Elizabeth, 'The Invisible Flaneur', *New Left Review* 191 (1992), pp. 90–110.

——, *Bohemians: The Glorious Outcasts* (London: I. B. Tauris, 2000).

Wilson, Michael, 'Portrait of the Artist as a Louis XIII Chair', in Gabriel Weisberg (ed.), *Montmartre and the Making of Mass Culture* (New Brunswick, NJ: University of Rutgers Press, 2001), pp. 180–204.

Wolff, Janet, 'The Invisible Flâneuse; Women and the Literature of Modernity', *Theory, Culture & Society* 2 (1985).

Wolin, Richard, 'Foucault's Aesthetic Decisionism', *Telos* 67 (1986), pp. 71–86.

Woodcock, George, *Anarchism. A History of Libertarian Ideas and Movements* (Peterborough, ON: Broadview, 2004).

Youngs, Tim, *Travel Writing in the Nineteenth Century: Filling the Blank Spaces* (London: Anthem, 2006).

Yusoff, Katherine, 'Biopolitical Economies and the Political Aesthetics of Climate Change', *Theory, Culture & Society,* 27 (2010), pp. 73–99.

Zola, Émile, *La Débâcle* (Oxford: Oxford University Press, 2000).

Zola, Émile, *L'Assommoir*, trans. Margaret Mauldon (Oxford: Oxford University Press, 1995).

Zukin, Sharon, *Loft Living: Culture and Capital in Urban Change* (London: Radius, 1988).

# Index

Adorno, Theodor 83, 85, 97, 160
aestheticization ix, xi, 122, 138, 163, 176
aesthetics of existence 9, 10, 177, 178,
 195; *see also* art of life
affect x–xii, xiv, 14, 19, 20, 23–6, 34,
 59, 66, 70, 73, 84–6, 97, 101,
 109–13, 121–3, 150, 162, 165,
 171, 174, 175, 179–82, 187,
 188, 190–92, 196–8; *see also*
 experience; life
affirmation 25, 64, 82–5, 111–12, 118–22,
 125, 136, 163–5, 174–6, 196
 affirmative pessimism 121–3
alcohol 2, 17, 32, 52
alienation 8, 14, 37–8, 41, 59, 68–71, 103,
 111, 118, 121–2 , 144–5, 158–60
*An Enemy of the People* 191
anarchism ix, x, xii, xiv, 4, 7, 23, 26, 29,
 31, 49, 55, 62, 64, 75, 77, 78,
 93, 103, 110, 114, 165, 169–194,
 196–8; *see also* authority,
 anarchism and
anti-authoritariansim 2, 23, 75; *see also*
 anarchism; authority, crisis of
Anti-Authoritarian International 75
*Anti-Concierge, see L'Anti-Concierge*
anti-museum 127, 130–33, 136; *see also*
 museum
Arendt, Hannah 54, 176, 187
art of life xii, 8–14, 21, 22, 66, 70, 177;
 *see also* aesthetics of existence
aura 58–62, 137
authority
 aesthetics of 23, 195–200; *see also* art
 of life
 anarchism and 169–93
 biopolitical 65, 66, 71, 184, 197
 crisis of xii, xiii, 2, 48, 75, 196
 experiential xi, xiii, xiv, 49, 58–70, 71,
 73, 85–6, 105, 110, 112, 121, 123,

146, 184–6, 197, 198; 200; *see also*
 authority, immanent
 experimental xi, xiii, 49, 55–8, 71;
 86–92; 178; *see also* experimental
 embodiment; Positivism
 humour and 23, 26, 78, 101, 105, 110,
 121, 125, 132, 170, 187
 immanent x, 22, 23, 49, 51;
 182–7, 195; *see also* authority,
 experiential
 traditional, *see* authority, transcendent
 transcendent xiii, 1, 9, 22, 49, 51–5,
 71, 105, 108, 170, 182, 195
 truth and 6, 14, 18, 22, 49, 123, 121,
 136, 137, 139, 142, 177, 178, 179,
 182, 184, 186, 187, 195, 197;
 *see also* parrhesia
 violence and 50, 75, 171, 187
avant-garde x–xiii, 4–7, 10–14, 18, 22,
 26, 78, 87, 97, 101, 102, 104, 107,
 110, 111, 114, 118, 120, 121, 122,
 141–2, 157, 160, 170, 171, 181,
 191–3; *see also* Dada; Situationists

bacteria 17, 57
Bain, Alexander 110, 187–8
Bakunin, Mikhail 162, 183
Barrucand, Victor 169, 182–4
Basilica de Sacré Cœur ix–x, 55, 105, 107
Bataille, Georges 24, 43
Baudelaire, Charles xiii, 7, 10, 29–31,
 34–41, 47, 65–70, 71, 73, 114,
 118–19, 141, 151, 157, 190, 195,
 196
beauty 22, 29, 32, 37–41, 44, 114, 126, 152
Belleville 51
Benjamin, Walter 10, 38, 59–62, 65,
 67–70, 75, 160, 199
Bergson, Henri 13, 84–5, 90, 110, 176, 193
Bernard, Claude 56